THE WONDERS OF BAMBOO REINFORCEMENT
GREEN SOLUTIONS

THE WONDERS OF BAMBOO REINFORCEMENT
GREEN SOLUTIONS

Dr. Shilpa Pankaj Kewate

Prof. (Dr.) B.K. Sarkar

NAMYA PRESS.com

Published by

Namya Press

USA: 10685-B Hazlehurst Dr. # 31013 Houston, Texas - 77043
Email: namyapress@gmail.com
Website: https://namyapress.com

Edition	:	First Published in 2022 (Namya Press)
Title	:	The Wonders of Bamboo Reinforcement: Green Solutions
Editors	:	Dr. Shilpa Pankaj Kewate, Prof (Dr.) B.K. Sarkar
ISBN	:	978-1-956861-17-4

Preface

It's with great happiness that we would like to acknowledge a great deal of people that get helped me extremely through the entire difficult, challenging, but a rewarding and interesting path towards some sort of Edited Book Without having their help and support, none of this work could have been possible.

Dr. Shilpa Pankaj Kewate, Prof (Dr.) B.K. Sarkar highly indebted to Dr. Vasudevan Pillai, Secretary and CEO, and Dr. Daphne Pillai, Joint Secretary and Chairperson, Management Board of Mahatma Education Society, for their kind support and encouragement.

Contents

Chapter - 01

Bamboo Reinforced Manhole Cover: An Initiative towards Sustainable Livelihood Development from Agriculture

Dr. Shilpa Kewate

Associate Professor, Department of Civil Engineering, Pillai HOC College of Engineering and Technology, Rasayani, Maharashtra, India. Email: shilpakewate@mes.ac.in,

ABSTRACT

The Livelihood status of farmers is quite pitiable in Vidarbha in Maharashtra state as extreme reliance on agriculture has worsened the situation. Farmers commit suicide due to their inability to repay loans mostly taken from banks to purchase expensive seeds and fertilizers. This motivates the author to start a venture to create a sustainable livelihood for the farmer, with the skill they have i.e., farming. Bamboo has potential both as an environmentally preferred material and means of employment generation. For the sustainable livelihood of farmers small-scale industries can be set up for the construction of bamboo reinforced concrete manhole cover. Bamboo is a naturally available resource in the Vidarbha in Maharashtra State. It can be used as an alternative to steel in a reinforced concrete manhole cover. Literature reveals that in recent years, steel prices have soared so replacement of steel reinforcement with alternative material is gaining immense importance. Bamboo is an alternative material for steel, which will have a tremendous economic advantage. The broad objective of this paper is to provide a preliminary contribution toward the sustainable livelihood of farmers through small scale bamboo reinforced concrete manhole cover construction industry with bamboo as an alternative to steel.

Keywords: *Sustainable, Environment, Bamboo, Manhole-cover, agriculture.*

INTRODUCTION

This Literature reveals that in recent years, steel prices have soared so replacement of steel reinforcement with alternative material is gaining immense importance (Nayak et al.2013).Bamboo is an alternative material for steel, which will have a tremendous economic advantage. In India, many open manholes have been death traps for pedestrians, especially during monsoon. Newly develop "Bamboo reinforced concrete manhole cover" would be cost-effective, environment friendly, and can be successfully utilized to avoid accidental death due to open manhole cover sewer system. The livelihood status of farmers is quite pitiable(Dongre and Deshmukh 2012). Therefore, an alternative employment option should be offered to farmers and their families. From the current scenario of farmers in the it is very clear that it is high time to start a venture to create a sustainable livelihood for them with the skill they have i.e., farming.

As there is enormous industrialization globally, the release of dangerous gases in the environment from various manufacturing sectors has imbalanced the ecosystem. And a cause for unfavourable conditions such as unpredictable rainfall, rise in sea level, global warming, acidification of oceans, earthquakes, etc. and also the crucial problem encountered by the construction sector is carbon emission from steel and cement manufacturing. The construction sector contributes up to 11% of carbon dioxide emissions globally.

For manufacturing 1 ton of steel, there is an emission of approximately 2 tons of carbon dioxide, and for manufacturing 1 ton of cement, there is an emission of approximately 1 ton of carbon dioxide in the environment (Bhalla 2008). Due to the increase of carbon footprint in the environment, the earth's temperature rises year by year. If this situation continues, it will cause an unfavorable environment for humans and other living organisms, and the sustainability of life will be an issue.

Concrete has good compressive strength but low tensile strength (falnu et al.2014). To overcome this condition in concrete structures, steel is introduced as reinforcement in concrete for providing the tensile strength required for the stability of the structure (Sutharsan et al.2020). However, some issues commonly faced with steel are increment in overall cost, use of the non-renewable resources for its manufacturing, polluting environment due to industrial processing, a significant contributor for the overall dead load of structures, etc. So there is a need to have an alternative and environmentally friendly material in construction that won't harm the environment, similar to steel, which can contribute to sustainable development. Steel reinforced manhole cover, and mild steel manhole cover has a good resale value due to which these manhole covers are in danger of being stolen, which may cause accidents due to open holes left on the road. Furthermore, it may harm the life of the public and animals walking on roads. Again, the open holes of the manhole may disturb the lives of residing people due to foul odours and arising to various diseases.

PROPOSED OUTCOMES

This Echo friendly product as shown in figure1 ('Bamboo reinforced manhole cover') would be developed and commercialized.

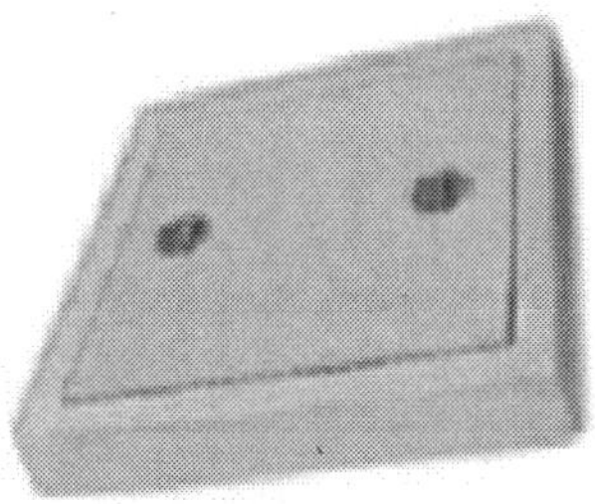

Figure 1: Bamboo reinforced manhole cover

1. The cost of construction of 'Bamboo reinforced manhole cover' may be reduced up to 50% than the conventional construction.
2. This research work will be surely helpful for the sustainable livelihood of farmers in the Vidarbha region through the small-scale manhole construction industry.
3. This proposal has the potential to employ all family members of farmers in processes like, acquiring material, buying bamboo, applying the waterproof coating, casting, manufacturing, promotion of a newly developed product, marketing, selling.

Executive Summary

Concrete manhole cover, prepared from environmentally friendly material bamboo as a replacement of steel would also help in reducing CO2 emission. Traditional manhole cover can be replaced by GREEN MANHOLE COVER at 50% cost. Bamboo has potential both as environmentally preferred material and means of employment generation. So, bamboo reinforcement technology can play important part in development of enterprises and transformation of rural Vidarbha region.

Opportunity

The product developed would be GREEN MANHOLE COVER (Bamboo reinforced concrete manhole cover), which is sellable in market. Construction industry is the main consumer of steel. From the literature survey it has been observed that availability of steel has become a major issue. Bamboo is naturally available resource in Vidarbha region. It can be used as alternative to steel in reinforced concrete manhole cover. Author's target market competition would with established manhole producing companies and target market would be government organization like Nagpur improvement trust (NIT), Nagpur municipal cooperation (NMC), MIDC industrial area around Nagpur. Targeted Primary customer Nagpur improvement trust (NIT) Nagpur municipal cooperation (NMC) MIDC industrial area around Nagpur. Nagpur improvement trust (NIT).Business plan opportunities are shown in figure 2.

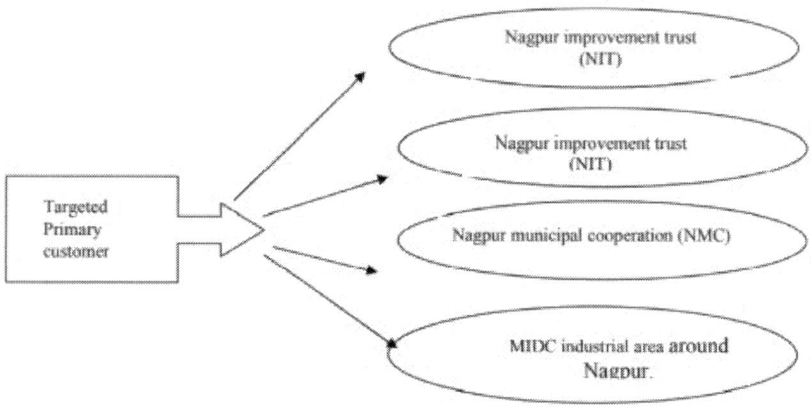

Figure 2: Opportunity for Business plan

Execution

For economic empowerment of farmer in Vidarbha region small scale industries can be set up for construction of manhole cover. Product developed can be sold in market thus revenue generated could be used for Self- sustainability of the project.

Financial Plan

For this project expected Recurring cost is Rs 92,000/- and Non-recurring cost is Rs. 5,000/. Complete budget details are given in table 1.

Table 1: Budget details

Sr.No.	Particulars/ Items	Total Cost (in INR)
1.	Recurring (in detail)	
	a)Consumables(Bamboo, cement ,sand, aggregates, admixtures, coal tar etc)	40,000
	b) Manufacturing	30,000
	c) Field visit and testing	20,000
	d) Miscellaneous	2000
2.	Non-recurring (in detail)	
	a) Fabrication of mold (shape Manhole cover)	5000
	Grand Total	INR 97,000/-

METHODOLOGY

Following methodologies will be adopted to achieve stated objectives.

- Bamboo cannot be used directly as reinforcement in concrete thus bamboo has to be properly selected, seasoned, and treated before using.
- Bamboo reinforced concrete manhole cover will be constructed tested and will be compared with a corresponding steel-reinforced manhole cover.
- Plan for sustainable livelihood of farmers in Vidarbha region in Maharashtra through small scale 'Bamboo reinforced concrete manhole cover 'construction industry with bamboo as an alternative to steel will be proposed.

The flowchart showing the methodology for the preparation process of a bamboo reinforced manhole cover is shown in Figure 3.

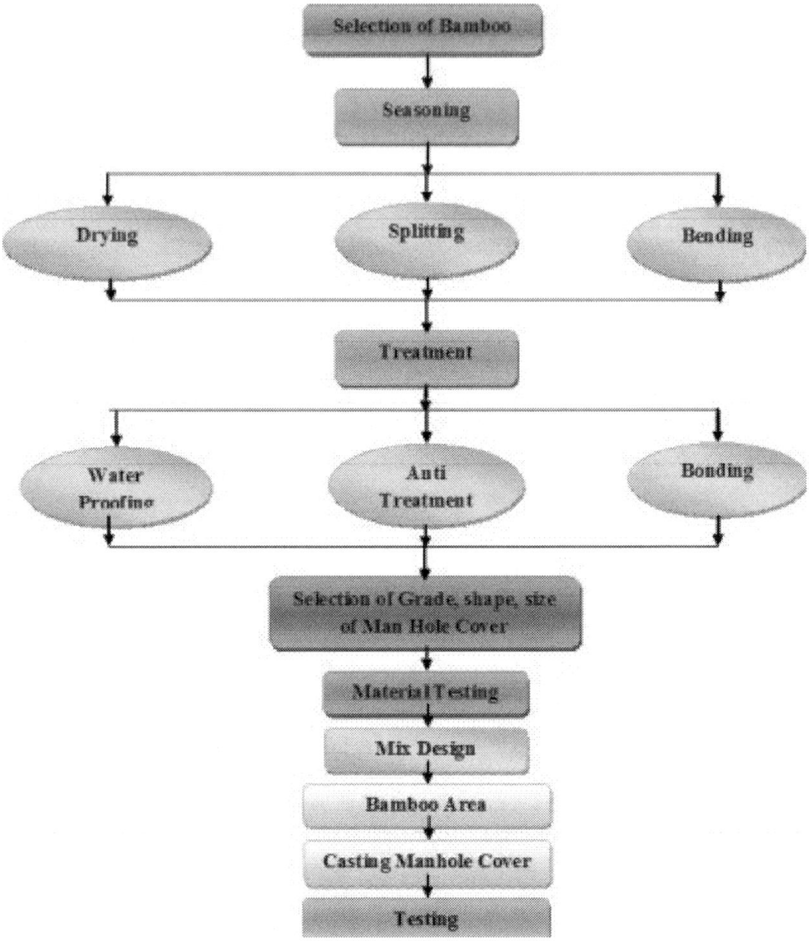

Figure 3: Flowchart of Methodology

CONCLUSION

Bamboo is an alternative material for steel, which will have a tremendous economic advantage. Newly develop "Bamboo reinforced concrete manhole cover" would be cost-effective, environment friendly, and can be successfully utilized to avoid accidental death due to open manhole cover sewer system. This alternative employment option should be offered to farmers and their families to start a venture to create a sustainable livelihood for them with the skill they have i.e., farming.

REFERENCES

[1] Bhalla, S. et al. (2008) "Bamboo as Green Alternative to Concrete and Steel for Modern Structures", *Journal of Environmental Research and Development*, 3(2), pp. 362-370.

[2] Dongre Amol R. and Deshmukh Pradeep R, "Farmer's suicides in the Vidarbha region of Maharashtra, India: a qualitative exploration of their causes" *Journal of Injury and Violence Research (JIVR)*, Volume 4 (1): 2–6.doi: 10.5249/jivr.v4i1.68 (Jan 2012).

[3] Fanlu Min, Zhanhu Yao, Teng Jiang,(2014) "Experimental and Numerical Study on Tensile Strength of Concrete under Different Strain Rates", *The Scientific World Journal*, vol. 2014, ArticleID 173531, 11 pages, 2014. https://doi.org/10.1155/2014/17 3531.

[4] Kaware Ajinkya,Awari U., Prof. Wakchaure M.,(2013), "Review Of Bamboo As Reinforcement Material In Concrete Structure," *International Journal of Innovative Research in Science, Engineering and Technology* Vol. 2, Issue 6.

[5] Nayak Anurag , Bajaj Arehant, Abhishek Jain, Apoorv Khandelwal, Hirdesh Tiwari, "Replacement of Steel by Bamboo Reinforcement." *IOSR Journal of Mechanical and Civil Engineering* (IOSR-JMCE) e-ISSN: 2278-1684,p-ISSN: 2320-334X, Volume 8, Issue 1 (Jul. - Aug. 2013), PP 50-61.

[6] Sutharsan R.,Ramprasanna S.(2020),"Experimental study on Bamboo as a reinforcing material in concrete", *AIP Conference Proceedings* 2204, 020024 (2020.)

CITATIONS

- Dr. Reni Francis ,2021107208, Granted, MAHATMA EDUCATION SOCIETY'S ," Analyzing the Role of Edu-2030 Model in Developing Competencies with respect to SDGs among Student Teachers" ,2021, Australia Patent Office.
- Francis, Reni; Cherian, Jaya ,2021105656, Granted, MAHATMA EDUCATION SOCIETY'S "Analysing the role of SHANTI Model of Teaching in fostering Peace among Secondary School Students" 2021, Australia Patent Office.
- Mr. Raju Narwade, Mr. Aditya Jadhav, Mr. Varad Salunkhe, Mr. Atharva Ugale, Ms. Smitha J, 202121039462 ,MAHATMA EDUCATION SOCIETY'S PILLAI HOC COLLEGE OF ENGINEERING AND TECHNOLOGY ," Method and Process of Manufacture P-Tiles using Waste Plastic and other Required Material" 2021,Indian Patent office.
- Dr. Lata Menon, Lalit Mahato, Yash Khopkar, Priya Sharma, Ms. Darshana Wajekar, Mr. Ravi Bari, 202121039911,MAHATMA EDUCATION SOCIETY'S PILLAI HOC COLLEGE OF ARTS, SCIENCE AND COMMERCE "FLSA-System: Fifth-level Authentication System (Mobile, Laptop, PC, Locker, Etc.)" 2021, Indian Patent office.
- Dr. Archana Bhagwat, Ms. Krutika Mahendra Malkar, Mr. Abhishek Jeevan Ingole, Ms. Srushti Satyawan Sawant, Dr. Vishakha Telgote, 202121040919, MAHATMA EDUCATION SOCIETY'S PILLAI HOC COLLEGE OF ARTS, SCIENCE AND COMMERCE, "Intelligent Process and Methods for Casein Determination in Cow-milk only" 2021, Indian Patent office.
- Dr. G. V. Patil , Sagar Tate, Pranav Gawand, Gaurav Patil, Mr. Hemant Patil, Ms. Ashwini Kadam, 202121042117, MAHATMA EDUCATION SOCIETY'S PILLAI HOC COLLEGE OF ENGINEERING AND TECHNOLOGY," Automatic Air Fill in Bike, Scooty and Car Tyres" 2021, Indian Patent office.
- Dr. Biplab Kumar Sarkar, US-17463679, MAHATMA EDUCATION SOCIETY'S PILLAI HOC COLLEGE OF ENGINEERING AND TECHNOLOGY "Advanced Kaplan Turbine." 2021, US-Patent Office.
- Mr. Vikas Suresh, Mr. Raju Narwade, Dr. G.R. Patil,202121043251, MAHATMA EDUCATION SOCIETY'S PILLAI HOC COLLEGE OF ENGINEERING AND TECHNOLOGY, "SMART CONSTRUCTION SAFETY HELMET: An Arduino-Based Risk, Salary Monitoring and Management Tool, 2021,Indian Patent Office.

- Ms. Neethumol K G , Roshini Singh, Prerna Tupe, Ravi Bari, 202121043253, MAHATMA EDUCATION SOCIETY'S PILLAI HOC COLLEGE OF ARTS, SCIENCE AND COMMERCE, " SV-Calculator: Scientific Voice Calculator" 2021, Indian Patent Office.
- Pragnesh Narendra Shah, 202121043844, MAHATMA EDUCATION SOCIETY'S, PILLAI COLLEGE OF ENGINEERING," Intelligent Electrical Vehicles (IEV) Charging (No need of charging stations), 2021, Indian Patent office.
- Pranav Umesh Parab, Prof(Dr.) Shilpa Kewate, 202121043843, MAHATMA EDUCATION SOCIETY'S, PILLAI HOC COLLEGE OF ENGINEERING AND TECHNOLOGY," Manhole Cover with Bamboo as Sustainable, Reinforcement in Light Duty Grade" 2021, Indian Patent Office.
- Munawira Kotyad, (Associate Professor) , Anish Sivarajan Kunnathuparambil, (Assistant Professor)202121044387, MAHATMA EDUCATION SOCIETY'S, PILLAI HOC COLLEGE OF ENGINEERING, " Gesture Recognition Glove Integration with Home Assistant Systems for Verbally Challenged Persons " 2021, Indian Patent Office.
- Dr. Jayanta Kumar Behera, Dr. Sapana Chilate ,202121047457, MAHATMA EDUCATION SOCIETY'S PILLAI HOC COLLEGE OF ARTS, SCIENCE AND COMMERCE, " Device to Detect the Fruit and Vegetable Classification and fruit (Apple) Disease" 2021, Indian Patent office.
- Ms. Neethumol K G , 202121043253, MAHATMA EDUCATION SOCIETY'S PILLAI HOC COLLEGE OF ARTS, SCIENCE AND COMMERCE RASAYANI TALUKA KHALAPUR DISTRICT RAIGAD MAHARASHTRA 410207, INDIA." SV-Calculator: Scientific Voice Calculator", 2021, Indian Patent Office.
- Dr. Divya Chirayil, Ms. Rupali Sathe, 202121047470, MAHATMA EDUCATION SOCIETY'S, PILLAI HOC COLLEGE OF ENGINEERING AND TECHNOLOGY," IoT-Kit to Track and Notify Doctor, Nurse and Medical Equipment" 2021, Indian Patent office.
- Gaurav Gyanendra Mishra, Akash Surendra Yadav, 202121046356, MAHATMA EDUCATION SOCIETY'S, PILLAI HOC COLLEGE OF ENGINEERING AND TECHNOLOGY," ECO-FRIENDLY POLYPROPYLENE PAVER BLOCK WITH SWAPPABLE ANTI SKID FILM, 2021, Indian Patent office.
- Dr. Nivedita Shreyans (Associate Professor), 202121046795, MAHATMA EDUCATION SOCIETY'S, PIMSR-PILLAI INSTITUTE OF MANAGEMENT STUDIES AND RESEARCH, "Method and Process MCQ Examination Management System, 2021, Indian Patent office.

- Dr. Bhavna Dave (Assistant Professor), Mr. Hardik Dave (Assistant Professor), 202121048707, MAHATMA EDUCATION SOCIETY'S," English Speaking Error Detector AI- Based Mobile Chip, 2021, Indian Patent office.
- Dr. Sally Enos, Rohan Enoss Mathew, 202121048708, MES' PILLAI COLLEGE OF EDUCATION AND RESEARCH, Plot No 1, Sector-8, Khanda Colony, New Panvel (W), "Portable Tool to hold the Umbrella Travelling in a Car.", 2021, Indian Patent office.
- Dr. Reni Francis – Principal, Reuben Francis ,Ryan Francis, 351149-001, Design Patent, MAHATMA EDUCATION SOCIETY'S, MES' PILLAI COLLEGE OF EDUCATION AND RESEARCH - Chembur Mumbai. "Trio pen pencil and eraser, 2021, Indian Patent office.

Chapter - 02

Bamboo Reinforcement: A Sustainable Technology for Energy and Environment Management

Pranav Umesh Parab[1], Dr. Shilpa Pankaj Kewate[2]

[1]PG Student, Department of Civil Engineering, Pillai College of Engineering and Technology, Rasayani 410207, Maharashtra, India. Email: pranavparab15041995@gmail.com

[2]Associate Professor, Department of Civil Engineering, Pillai College of Engineering and Technology, Rasayani 410207, Maharashtra, India. Email: shilpakewate@mes.ac.in

ABSTRACT

One of the major environmental problem faced by the construction industry is carbon emission from steel and cement manufacturing. Due to the increment in carbon footprint, the earth's temperature is rising year by year. If the situation continues humans, as well as other living organisms on earth, may become extinct due to unfavorable environmental conditions. This study focuses on the reduction of the carbon footprint from steel manufacturing by replacing steel reinforcement with bamboo which is an environmentally friendly and renewable material. Using bamboo as reinforcement the process of selection, seasoning and treatment are to be carried out on bamboo before using it, in concrete for archiving best results. The requirement of water-cement for bamboo reinforced concrete and the percentage of bamboo to be used in different structural members is studied. The strength results of various bamboo reinforced structural members are shown. Using bamboo reinforcement cost can be reduced to half or more as compared to steel reinforcement with similar strength results. The study also concluded that bamboo reinforcement can be used for lightweight structures which could majorly contribute to rural development. Also, it will help in achieving sustainable development without endangering the lives of living organisms on earth.

Keywords: *Bamboo reinforcement, Carbon footprint, Steel reinforcement, Sustainable development.*

INTRODUCTION

General

As the vast industrial development pollution of the environment has reached its peak level. This has disturbed the ecological atmosphere and a cause of environmental changes resulting in acidification of oceans, rising sea level, uneven rainfall, droughts, global warming, earthquakes, tsunami, etc. The major issue faced by the construction industry is carbon emission from steel and cement manufacturing. In the manufacturing of 1 ton of steel, there is an emission of approximately 2 tons of carbon dioxide and for 1 ton of cement, there is approximately 1 ton of carbon dioxide emitted in the environment (Suresh et al, 2008). As a result of the increase of carbon footprint in the environment the earth's temperature is rising year by year. If this condition remains the same then it will make an uncomfortable surrounding for humans as well as other living organisms existing on earth.

Concrete has good compression strength but low tensile strength. To overcome this condition in concrete structures, steel is introduced as reinforcement in the concrete for providing the tensile strength required for stabilizing the structure. Some commonly faced problems with steel are increment in the overall cost of a structure, use of a large amount of non-renewable resources for its manufacturing, polluting the environment from its manufacturing process, a major contributor to the overall dead load of structures, etc. So there is a requirement to have a substitute and eco-friendly material in construction that won't damage the environment gives similar strength as steel and can contribute towards sustainable development.

Bamboo

Bamboo is a part of the grass family *Poaceae* belongs to the sub-family of *Bambusoideae.* It is a composite material of long and parallel fiber cellulose fibres embedded in a ligneous matrix (Khosrow, 2004). It is a fast-growing woody grass that can grow fully in 3-5 years with less water requirement and can sustain mostly in any type of soil and weather conditions. There are an abundant number of species available if suitable species have to be taken for cultivation according to the regional atmosphere. The carbon sink by 5-year-old bamboo every year is 18 to 23 tons per acre. Table 1 shows the various mechanical properties of bamboo.

Table 1: Mechanical properties of bamboo (Zhao, Zhau and Bu, 2012).

Mechanical Properties	Value (SI)
Allowable curve compressive stress	65 Mpa
Allowable axial compressive stress	50 Mpa
Ultimate tensile stress	600 Mpa
Allowable tensile stress	400 Mpa
Allowable shearing stress	4.5 Mpa
Allowable bond stress	0.35 Mpa
Modulus of elasticity	14000 Mpa
Density	700 kg/m3

Advantages of Bamboo

Bamboo is a natural, renewable, and biodegradable material. Its tensile strength to specific weight ratio is high, also it possesses both tensile and compressive strength (Anurag et al, 2013). The cost of bamboo is cheaper as its availability is abundant in rural as well as the urban region. It can prevent pollution by absorbing a large amount of nitrogen from wastewater (Akash, Anil, and Vikas, 2014).

Disadvantages of bamboo.

The bamboo tensile strength is relatively very lesser than steel (Khan, 2014). Its water absorption capacity is very high which affects the dimension and develops low bond strength with concrete (Dinesh et al, 2014). The structure of bamboo is irregular thus the strength may vary drastically. As it is a woody grass it can be attacked by termite and decay (Shakeel, Altamash and Hina, 2014).

PROCEDURE FOR USING BAMBOO AS REINFORCEMENT

General

According to literature studies conducted the bamboo is incapable of direct use as reinforcement in concrete thus it has to be properly selected, seasoned, and treated before use. Figure 1 shows the flowchart showing the procedure for using bamboo reinforcement.

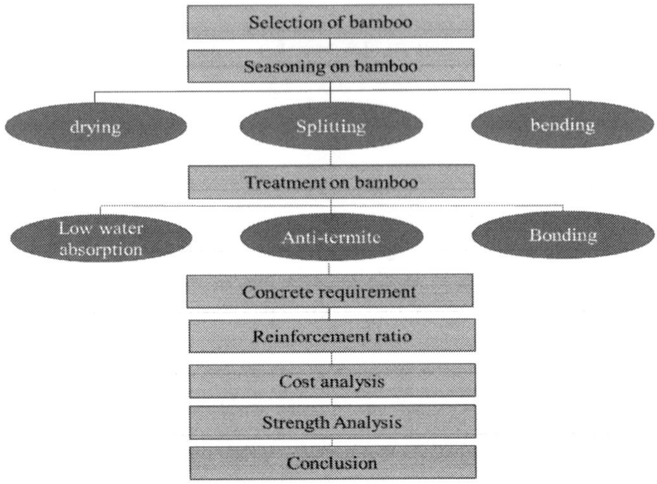

Figure 1: Flowchart of procedure for using bamboo reinforcement.

Selecting

Bamboo showing brown color should be used which ensures that the bamboo is at least three years old. Selecting the longest and largest diameter of bamboo culm ensures its full growth. Do not use whole culms of green, unseasoned bamboo as the culms are generally weaker (Francis and Paul, 1966).

Seasoning

The harvested bamboo has to be dried for three to four weeks to reduce the moisture content of bamboo fibers. Reducing the moisture content increases the overall strength of bamboo. The complete culm of bamboo cannot be used as reinforcement, so the whole culm has to be split vertically with a sharp knife. Splitting will result in continuous fibers and a nearly straight section as per the desired size (Francis and Paul, 1966).

Treatment

As bamboo has a very high water-absorbing capacity hence it has to be coated with coal tar, grease, or epoxy coating and a sprinkling of sand which improves bonding with concrete, allow this coating to dry at least 24 to 48 hours before using as reinforcement. Bamboo can be attacked by termite so it has to be treated with lethal solution by soaking or applying a coat on it (Shakeel, Altamash and Hina, 2014; Dinesh et al., 2014).

Concrete Requirement for Bamboo Reinforcement

The water-cement ratio of bamboo reinforced concrete should be as low as possible as bamboo has a high can water-absorbing capacity so it may absorb water from concrete (Francis and Paul, 1966; Akash, Anil, and Vikas, 2014). Chemicals such as super-plasticizers or water-reducing admixtures can be used to improve the workability of concrete.

Reinforcement Ratio (Area of Bamboo Reinforcement)

As bamboo tensile strength is low as compared to steel reinforcement the area of bamboo reinforcement required to satisfy the tensile requirement is more than steel. The required area of bamboo reinforcement is as follows:

- **Beam:** the area of bamboo reinforcement is taken as 3 – 4 % area of the cross-section.
- **Column:** the area of bamboo reinforcement is taken as 3 – 4 % area of the cross-section.
- **Slab:** the area of bamboo reinforcement is taken as 4 times the area of steel reinforcement (Zhao, Zhau and Bu, 2012).

STRENGTH OF BAMBOO REINFORCED CONCRETE STRUCTURES

Bamboo Reinforced Beam

The testing of bamboo reinforced beams of size 750x150x150mm. Three types of beams were investigated: plain concrete beam, singly bamboo reinforced, and doubly bamboo reinforced beam which was tested for flexural strength. From the test results, the flexural strength of bamboo reinforced beam is nearly double the as compared to plain beam without reinforcement (Shakeel, Altamash and Hina, 2014). Figure 2 shows the test setup for testing bamboo reinforced beams for flexural test and Table 2 shows the test results.

Table 2: Test results for bamboo reinforced beam (Shakeel, Altamash and Hina, 2014)

Types of beam	Ultimate Load (KN)	Maximum Deflection ($x10^{-2}$)mm	Flexural Strength (MPa)
Plain concrete beam	3.0	56	48
Singly bamboo reinforced beam	3.5	54	56
Doubly bamboo reinforced concrete beam	5.0	98	80

Figure 2: Test setup for bamboo reinforced beam (Shakeel, Altamash and Hina, 2014)

Bamboo Reinforced Column

The bamboo reinforced columns of various shapes by varying the percentage of reinforcement from 2.5 % to 4 % at an increment of 0.5%. the rectangular column of dimension 230x150x750 mm, square column of dimension 150x150x750 mm and circular column of 230 mm diameter and 750 mm length with varying percentage of bamboo were investigated for compression test in UTM. From the test results, the square column of 3.5% bamboo has taken the maximum load of 196.5KN. The rectangular column of 2.5% bamboo has taken the maximum load of 317KN. The circular column of 3% bamboo has taken the maximum load of 366KN (Ajinkya, Awari and Wakchaure, 2013). Figure 3 shows the test setup for testing bamboo reinforced columns in UTM. Table 3 shows the test results of the bamboo reinforced column.

Table 3: Test results of bamboo reinforced column. (Ajinkya, Awari and Wakchaure, 2013)

The maximum load carried by	square	rectangular	circular
for 2.5% bamboo reinforcement	192.9kN	317kN	329kN
for 3% bamboo reinforcement	170kN	280kN,	366kN
for 3.5% bamboo reinforcement	196.5kN	290kN	360kN
for 4% bamboo reinforcement	185kN	285kN	340kN

Figure 3: Test setup for the bamboo reinforced column. (Ajinkya, Awari and Wakchaure, 2013)

Bamboo Reinforced Slab

The experimental testing was done on the bamboo reinforced slab of 600x600x125 mm and tested the specimen for the flexural test in UTM. From the test results, the maximum load taken by bamboo reinforced slab is 16.2 KN (Dinesh et al., 2014). Figure 4 shows the test setup for the bamboo reinforced slab. Table 4 shows the details of slab specimens.

Table 4: Details of the bamboo reinforced slab. (Dinesh et al., 2014, p. 129)

Particulars	Values
Specimen Type	Bamboo Reinforced Slab
Concrete Grade	M20
Size of slab column	600x600x120 mm
Effective depth	99 mm
Effective span	490 mm
Reinforcement	Bamboo splints of Dendrocalmus Strictus
Area of reinforcement along span	1154.03Sqmm
Tensile strength of reinforcement	95.81MPa
Ultimate Load	16.200 KN
E_b	18600 MPa
E_c	22361 MPa

Figure 4: Test setup for the bamboo reinforced slab. (Dinesh et al., 2014, p. 129)

COST ANALYSIS OF THE BAMBOO REINFORCED CONCRETE STRUCTURE

The Cost Analysis of Slab, Beam, Column, and Footing

The cost analysis and comparing the cost of steel and 3% area of bamboo reinforcement (untreated) in the slab, beam, column, and footing. Approximately bamboo reinforcement is three times cheaper as compared to steel reinforcement (Anurag et al., 2013). Table 5 shows the cost analysis of steel and bamboo reinforcement.

Table 5: Cost analysis of steel and bamboo reinforcement. (Anurag et al., 2013)

Item	Steel (Rs.)	Bamboo (Rs.)
Slab	20371	9146
Beam 1	5008	640
Beam 2	4162	454
Column & Footing	5110	530
Total	Rs. 33,072	Rs.10,770

EXAMPLES OF BAMBOO REINFORCED CONSTRUCTION

Rigid Pavement Using Bamboo (Akash, Anil, and Vikas, 2014)

Construction of low volume road pavement using bamboo reinforcement. The cost of construction was reduced to more than half of rigid pavement. Figure 5 shows bamboo reinforced low volume pavement construction.

Figure 5: Low volume rigid pavement construction using bamboo (Akash, Anil, and Vikas, 2014)

Bungalow Construction using Bamboo (A Bamboo House, 2017)

The Bungalow owned by George P J is located near Karapuzha dam in Wayanad district, Kerala. This construction is done with the use of Ferro-cement using bamboo reinforcement. Figure 6 shows the Ferro-cement bungalow during construction and post-construction using bamboo reinforcement.

Figure 6a: Ferro cement Bungalow in the construction phase (A Bamboo House, 2017).

Figure 6b: Ferro cement Bungalow post-construction (A Bamboo House, 2017).

CONCLUSIONS

The various conclusion drawn from this study are as follows:

1. Bamboo has a very high water absorption capacity due to which it can swell and decay, so proper seasoning and treatment have to be done before using it as reinforcement in concrete. (Francis and Paul, 1966; Ajinkya, Awari and Wakchaure, 2013)
2. Bamboo reinforcement can reduce the cost of the structure by approximately half or more as compared to steel reinforcement. (Akash, Anil, and Vikas, 2014; Anurag et al., 2013, p. 59)
3. Tensile strength of bamboo is good and can be used as reinforcement in R.C.C structure for lightweight structure preferably single storey structure. (Anurag et al., 2013, p. 59)
4. Using bamboo reinforcement there can be a reduction in carbon footprint from the environment and can also lower the effect of global warming. (Suresh et al., 2008, p. 363)
5. Bamboo reinforcement can surely contribute majorly to rural and sustainable development by promoting green technology without endangering the lives of living organisms on earth. (Suresh et al., 2008, p. 369)

REFERENCES

[1] A Bamboo House, 2017, Mastercraft, Episode 48 Mathrubhumi News, 11 November 2017. Available at: https://youtu.be/MX2Dd8QLZ18

[2] Ajinkya K, Prof. U.R.Awari, Prof. M.R.Wakchaure, 2013, Analysis Of Bamboo Reinforced Concrete Column, International Journal of Innovative Research in Science, Engineering and Technology Vol. 2, Issue 6, June, pp. 2450-2459.

[3] Suresh B, Supratic G, Puttagunta S and Rupali S, 2008, Bamboo as Green Alternative to Concrete and Steel For Modern Structures, Journal of Environmental Research And Development Vol. 3 No. 2, pp. 362-370.

[4] Dinesh Bhonde, P. B. Nagarnaik, D. K. Parbat, U. P. Waghe, 2014, "Experimental Investigation of Bamboo Reinforced Concrete Slab," American Journal of Engineering Research, Volume-03, Issue-01, pp. 128-131.

[5] Francis EB and Paul JR, 1966, Bamboo Reinforced Concrete Construction, U. S. Naval Civil Engineering Labaoratory Port Hueneme, California.

[6] Dr. Shakeel A, Altamash R, and Hina G, 2014, Mechanical Properties of Bamboo Fibre Reinforced Concrete, 2nd International Conference on Research in Science, Engineering and Technology (ICRSET 2014), Dubai. Available at: http://dx.doi.org/10.15242/IIE.E0314522

[7] Khosrow G, 2005, Bamboo as reinforcement in structural concrete elements, Cement & Concrete Composites, pp. 637–649.

[8] Efe I, Christopher F, Olamiposi K and Otu E, 2015, Structural behaviour of bamboo-reinforced foamed concrete slab containing polyvinyl wastes (PW) as partial replacement of fine aggregate, Journal of King Saud University – Engineering Sciences, 29, pp. 345-355.

[9] Karthik S, Rao P and Awoyera P, 2016, Strength properties of bamboo and steel reinforced concrete containing manufactured sand and mineral admixtures, Journal of King Saud University - Engineering Sciences. Available at: http://dx.doi.org/10.1016/j.jksues.2016.12.003

[10] IK Khan, 2014, Bamboo Sticks as a Substitute of Steel Reinforcement in Slab, International Journal of Engineering and Management Research, Volume-4, Issue-2, pp. 123-126.

[11] Maruthupandian G. Saravanan R, Suresh KS and Sivakumar BG (2016), A Study on Bamboo Reinforced Concrete Slabs, Journal of Chemical and Pharmaceutical Sciences Volume 9, Issue 2, pp. 978-980.

[12] Anurag N, Arehant SB, Abhishek J, Apoorv K and Hirdesh T, 2013, Replacement of Steel by Bamboo Reinforcement, IOSR Journal of Mechanical and Civil Engineering, Volume 8, Issue 1, pp. 50-61.

[13] Ogunbiyi MA, Olawale SO, Tudjegbe OE and Akinola SR, 2015, Comparative Analysis of the Tensile Strength of Bamboo and Reinforcement Steel Bars as Structural Member in Building Construction, International Journal Of Scientific & Technology Research, Volume 4, Issue 11, pp. 47-52.

[14] Shweta P and Shruti M, 2014, Bamboo as a Cost Effective Building Material for Rural Construction, Journal of Civil Engineering and Environmental Technology, Volume 1, Number 6, pp. 35-40.

[15] Akash P, Anil KC and Vikas K, 2014, Use of Bamboo in Low Volume Rigid Pavement as Reinforced Material: A Review, Journal of Civil Engineering and Environmental Technology, Volume 1, Number 3 pp. 14-16

[16] Rao, P. and Karthik, S. 2018, Investigation on Flexural Behaviour of Beam with Bamboo as Main Rebars, International Journal of Recent Technology and Engineering, volume 7, pp. 173-177.

[17] Harish S, N.V. Vamsi Krishna Togati and I.V. Ramana Reddy, 2012, "Investigation on Properties of Bamboo as Reinforcing Material In Concrete," International Journal of Engineering Research and Applications, Vol. 2, Issue 1, pp. 77-83.

[18] Srimathi, S. et al. (2016), A Review of Bamboo as a Reinforcement Material in Modern Construction, International Journal of Science Technology & Engineering, 3(5), pp. 232-235

[19] Zhao W, Zhou J and B Guobin, 2012, Application Technology of Bamboo Reinforced Concrete in Building, Applied Mechanics and Materials, Volume 195-196, pp 297-302

CITATIONS

- Dr. Reni Francis – Principal, Reuben Francis,Ryan Francis 351150-001, Design Patent, MAHATMA EDUCATION SOCIETY'S, MES' PILLAI COLLEGE OF EDUCATION AND RESEARCH- Chembur Mumbai., "All in One Geometry Set," 2021, Indian Patent office.
- MAHATMA EDUCATION SOCIETY 'S, 511510-001, MES's Pillai College of Education and Research, Sector-8, Khanda Colony, New Panvel East, Navi Mumbai, Maharashtra 410207, India., "MES SPORK" 2021, Indian Patent Office.
- Dr. Lata Menon, Mr. Ravi Bari, Ms. Neethumol K G, 202121050104, MES's PILLAI HOC COLLEGE OF ENGINEERING &TECHNOLOGY, RASAYANI, " Bitcoin Price Quick Prediction Using AI, DL and ML", 2021, Indian Patent Office.
- Dr. Divya Chirayil, Ms. Rupali Sathe, Abhishek Jain, 202141050116, MES's PILLAI HOC COLLEGE OF ENGINEERING &TECHNOLOGY, RASAYANI, "Emotion Recognition and notification using Machine Learning and AI" 2021, Indian Patent Office.
- Dr. B.K.Sarkar, 20211105174, MAHATMA EDUCATION SOCIETY'S, PILLAI HOC COLLEGE OF ENGINEERING AND TECHNOLOGY, "Advanced Branding in the Marketing of Unilever Products using ML" 2021, Indian Patent office.
- Dr. Sandeep S. Shinde, Assist. Professor, Dr. Sejal Rohit Tambake, 202121052143, (O)MAHATMA EDUCATION SOCIETY, S VIDYADHIRAJA COLLEGE OF PHYSICAL EDUCATION AND RESEARCH, "Break in Chant detector of a Raider Playing Kabaddi" 2021, Indian Patent office.
- Dr. Mansi Subhedar (Associate Professor), Mr. Upendra Patil (Assistant Professor),202121052142, MAHATMA EDUCATION SOCIETY'S, PILLAI HOC COLLEGE OF ENGINEERING AND TECHNOLOGY," Light Weight Transform Domain IMAGE Steganography Using 32-bit ARM Cortex-M4 "2021, Indian Patent office.

Chapter - 03

Design of Cost-Effective Bamboo Reinforced Manhole Cover: A Step towards Sustainable Development

Dr. Shilpa Kewate[1], Ms. Manisha Jamgade[2], Ms. Madhulika Sinha[3]

[1]Associate Professor, Department of Civil Engineering, Pillai HOC College of Engineering and Technology, Rasayani 410207, Maharashtra,India. Email: shilpakewate@mes.ac.in

[2]Associate Professor, Department of Civil Engineering, Pillai HOC College of Engineering and Technology, Rasayani 410207, Maharashtra,India. Email: mjamgade@mes.ac.in

[3]Associate Professor, Department of Civil Engineering, Pillai HOC College of Engineering and Technology, Rasayani 410207, Maharashtra,India. Email: madhulikas@mes.ac.in

ABSTRACT

In India, many open Manholes have been death traps for pedestrians, especially during monsoon. Steel reinforced manhole cover and mild steel manhole cover have a good resale value due to which these Manhole covers are in danger of being stolen it may cause accidents due to open Manhole left on the road. This may harm the lives of the public and animals walking on roads. This research work aims to design an environment-friendly sustainable manhole cover by using bamboo as an alternative material to steel. This newly developed product is cost-effective. In this paper, experimental testing for load resisting capacity as per IS 12592:2002 of a bamboo reinforced Manhole cover and steel-reinforced Manhole cover is carried out. Also, cost-effectiveness between both the Manhole cover is presented in this paper. This research work approaches to contribute toward sustainable development of green alternative material to steel.

Keywords: *Design, cost-effectiveness, Bamboo reinforcement; Sustainable development.*

INTRODUCTION

The Manhole cover is the lid of the Manhole openings to a sewer system or inspection chamber (BS 497- 2006). Generally, Manhole covers are located on pavements and roads. It should be strong and durable to resist the load and impact of pedestrians and vehicles (ASTM C478-2008). These covers are normally made up of cast iron and concrete to bear the large loads (BS 7903 -2012).

The purpose of the Manhole cover is to avoid accidentally falling anyone or anything into the Manhole cover. And the Manhole cover allows the authorized person for repair and maintenance work. Theft of manhole cover for scrap value is the common issue associated with Manhole covers. It causes road accidents and several health hazards.

Steel reinforced Manhole cover and mild steel Manhole cover have a good resale value due to which these Manhole covers are at risk of being stolen which may cause accidents due to open holes left on the road (Salha and Hatow 2009).In 2012, the New York Times reported that based on the existing commodity prices for iron and steel, a stolen Manhole cover might fetch more than $30, but it costs the municipality about $200 to bring back each. Governments of Asian Countries like India and China have taken the initiative to replace the steel Manhole covers with concrete covers. But the thefts persisted since concrete covers were also stolen for the rebar inside the concrete (OÑA Mónica et al.2015).The open holes of a Manhole cover may disturb the lives of residing people due to bad odor and arises various diseases. Fig 1 show an accident caused due to missing Manhole cover on the road.

Figure 1: Accident due to missing manhole covers (Deshmukh, 2014).

Many researchers are working on the possibility of the use of bamboo as a structural element. Bamboo is the fastest-growing plant. It has enormous economic potential. From ancient times bamboo is used in the construction of bridges and houses. This research work approaches to design cost-effective Bamboo reinforced manhole cover which would contribute to sustainable development and also green alternative material to steel.

The novelty of this research work is about a totally newly developed product, 'Bamboo Reinforced Manhole Cover'. Bamboo is first time used as a substitute material to steel in the production of a Manhole cover. Even though many researchers have work on the use of bamboo as a substitute material in different structures but it is used first time in a Manhole cover. Bamboo has the potential of both environmental benefits as well as means of employment generation.

Types of Manhole Cover

- Fiber-reinforced plastic
- Glass-reinforced plastic
- Reinforced cement concrete
- Cast iron
- Steel fiber reinforced concrete
- Ductile iron.

Grades of Manhole Cover

The various grades of manhole cover, their grade, and use purpose is shown in Table 1.

Table 1: Grade of manhole cover (Source; IS 12592:2002)

Grade	Grade Designation (Load resist capacity)	Use/ purpose
Light Duty	LD-2.5 (25KN)	Residential and institutional Complexes/ areas with pedestrian, inspection chambers
Medium Duty	MD-10 (100KN)	Service lanes/roads, on pavements used under medium duty vehicular traffic, car parking areas
Heavy Duty	HD-20 (200KN)	Institutional/commercial areas/ carriageways/city trunk roads /bus terminals with heavy-duty vehicular traffic of wheel load between 50 to 100KN and parking areas and manhole chambers for road pavement

Grade	Grade Designation (Load resist capacity)	Use/ purpose
Extra Heavy Duty	EHD-35 (350KN)	Carriageways in commercial/industrial/ port areas/ warehouses/ godowns where frequent loading and unloading of trucks/ trailers, slow to fast-moving vehicular traffic for wheel loads up to 115 KN

BAMBOO

The fastest-growing plant on the earth is Bamboo. On degraded land also bamboo can be grown and harvested rapidly. Bamboo possesses high strength to weight ratio and tensile strength. Bamboo has the potential to use as a reinforcement material (Weifeng et al. 2012). Bamboo does not have natural durability. Durability of bamboo can be enhanced by soaking, drying and applying water proof coating and sand (Muhtar 2020).

(Ogunbiyi et al. 2015) has conducted an experimental investigation on tensile strength test on a variety of diameter sizes of steel and bamboo bars. Tensile results show that as the size of bamboo bars increases the tensile strength also increases.

Shweta Patil et al. (2014) have studied bamboo which can be a suitable replacement for steel. Bamboo can play a very important role in developing rural areas due to its cost-effectiveness. Also, the author has shown the advantages, disadvantages, treatments, and mechanical properties of bamboo.

I.K. Khan (2014) has conducted an experimental investigation on a bamboo reinforced slab of size 1000x1000x50 mm with different cross-sectional shapes of bamboo. Four different types of the slab were cast with mild steel-reinforced, circular bamboo reinforced, square bamboo reinforced, and triangular bamboo reinforced; these slabs were tested for single point load on center with all sides simply supported. From the test results, the load-resisting ability of the square bamboo reinforced slab is 25% as compared to the mild steel slab.

Ahmad et al. (2014) has investigated bamboo fibers in concrete cubes and bamboo reinforced beams of size 750x150x150mm. For concrete cubes, 1% bamboo fibers with varying sizes were tested for compression test. Three types of beams were cast plain concrete, singly bamboo reinforced and doubly bamboo reinforced beam which was tested for flexural strength. The test results of the compression test on bamboo fibers cubes show double the strength as compared to plain concrete cubes. From the flexural test results, the doubly reinforced bamboo beam has shown the best results as compared to others.

EXPERIMENTAL PROGRAM

Literature survey reveals that bamboo cannot be used directly as reinforcement in concrete thus bamboo has to be properly selected, seasoned and treated before using. Also the flowchart showing the experimental program for the preparation process of bamboo reinforced manhole cover is shown in fig 2.

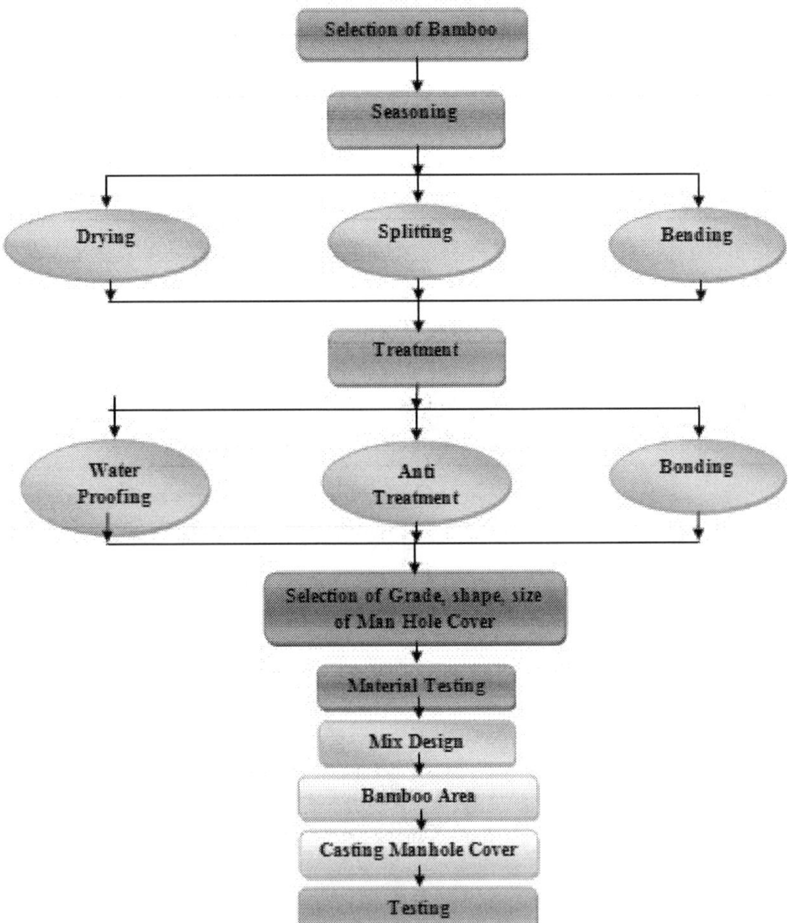

Figure 2: Flowchart of experimental work

The Casting of the Manhole Cover

According to the literature, survey bamboo cannot be used directly as reinforcement in concrete thus bamboo has to be dried to reduce moisture content, then after spitted vertically into strips form as per desired dimension (area) the cross-sectional area of single bamboo strips taken as 78.5mm2. Strips are then coated with coal tar and sprinkled sand on them which are well dried for 48hours. This makes the bamboo withstand water absorption and termites and a sprinkling of sand makes a good bond with concrete. Bamboo mesh tied with binding wire and then cast with M30 grade concrete. Fig 3 shows an experimental set up for testing of Manhole cover.

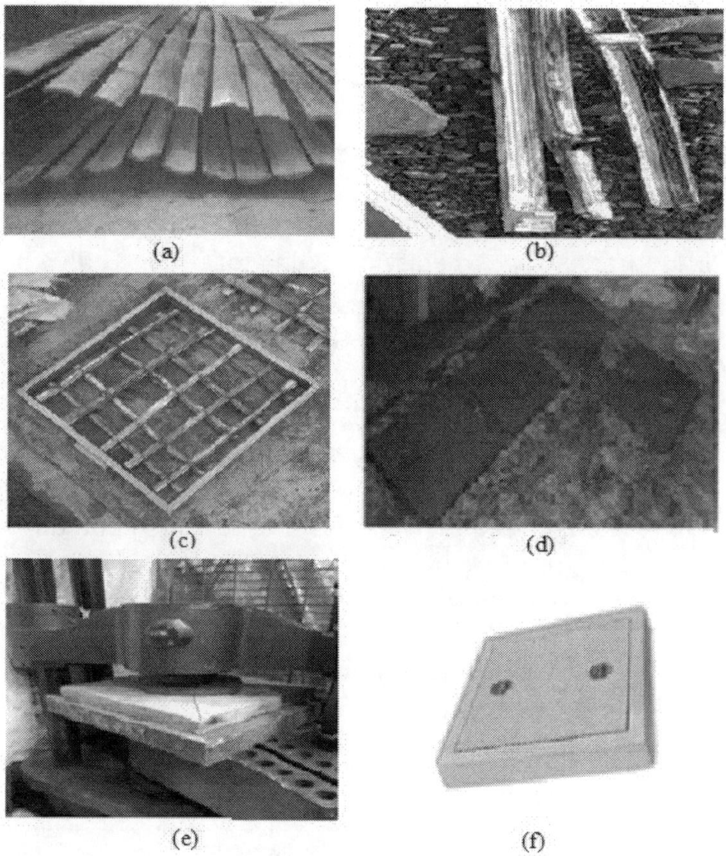

Figure 3: Experimental Setup (a) Splitting of bamboo (b) Bamboo strips coated with coal tar (c) Bamboo mesh for manhole cover (d) Casting of Manhole cover (e) Test setup for testing of Manhole cover in UTM (f) Final product

RESULTS

Load Resisting Capacity of Manhole Cover

Load test on Manhole cover is done in Universal testing machine (UTM) machine in this test application of target load is considered with the use of 300mm diameter circular block (plate) positioned on a manhole cover at its geometrical center and load application is done as per IS code 12592:2002 recommendation. As per IS 12592:2002, the minimum load resisting capacity of light-duty Manhole cover should be 25KN and heavy-duty Manhole cover should be 350 kN. Table 4.1 shows the load resisting capacity of Manhole cover.

In experimental work, the four cases of Manhole cover are considered.

1. Case 1: light-duty steel reinforced manhole cover of size 510x510x50 mm in which three samples were cast and tested for load resisting capacity and also for other analysis.
2. Case 2: light-duty 3% bamboo reinforced Manhole cover of size 510x510x50 mm in which three samples were cast and tested for load resisting capacity and also for other analysis.
3. Case 3: Extra heavy-duty steel reinforced Manhole cover of size 600x600x100 mm in which three samples were cast and tested for load resisting capacity and also for other analysis.
4. Case 4: Extra heavy-duty 3% bamboo reinforced Manhole cover of size 600x600x100 mm in which three samples were cast and tested for load resisting capacity and also for other analysis.

Table 4.1: Load test results (load resisting capacity) of manhole cover

Particular	Dimension (mm)	Reinforcement	Load (KN)	Average load (KN)
Case 1- light duty steel reinforced Manhole cover	510x510x50	Steel	25	25
	510x510x50	Steel	25	
	510x510x50	steel	25	
Case 2 - light duty 3% bamboo reinforced Manhole cover	510x510x50	3% Bamboo	18	20
	510x510x50	3% bamboo	19	
	510x510x50	3% bamboo	23	

Particular	Dimension (mm)	Reinforcement	Load (KN)	Average load (KN)
Case 3- Extra heavy duty steel reinforced Manhole cover	600x600x100	Steel	350	
	600x600x100	Steel	350	
	600x600x100	Steel	350	350
Case 4- Extra heavy duty 3% bamboo reinforced Manhole cover	600x600x100	3% bamboo	62	65
	600x600x100	3% bamboo	65	
	600x600x100	3% bamboo	68	

Cost Effectiveness

The total cost of case 2 bamboo reinforced manhole cover is 36% less than as compared to case 1 steel reinforced cover.

Case 1 light- duty steel reinforced manhole cover

Table 4.2. shows the cost analysis of case 1, size 510x510x50mm light duty steel reinforcement Manhole cover

Table 4.2: Cost analysis of case 1- light-duty steel reinforced manhole cover

Case 1- light -duty steel reinforced Manhole cover	Concrete M30 (cum) (Rs. 5000/ cum)	Reinforce-ment (kg.) (Rs. 50000/ ton Steel)	Treat-ment coal tar	Total (Rs.)
Quantity	0.013	2.01	-	
Price Rs.	65	100.5	0	165.5

Case 2 light -duty 3% bamboo reinforcement

Table 4.3 shows the cost analysis of case 2, size 510x510x50mm light duty bamboo reinforcement Manhole cover

Table 4.3: Cost analysis of light-duty 3% bamboo reinforced Manhole cover

Case 2 - light -duty 3% bamboo reinforced Manhole cover	Concrete M30 (cum) (Rs. 5000/ cum)	Reinforcement (Rs. 100 per bamboo)	Treatment coal tar	Total (Rs.)
Quantity	0.013	Less than 1/3 bamboo used	Rs. 10	
Price (Rs.)	65	30	10	105

From cost analysis given in table 1 table 2 it is observed that the total cost of a light-duty bamboo reinforced manhole cover (Newly developed Manhole cover) is 36% cheaper as compared to a light-duty steel reinforced manhole cover (existing Manhole cover). This newly developed product is cost-effective and it will contribute toward sustainable development of green alternative material to steel.However, the further experimental investigation should be conducted to enhance the load-carrying capacity of the bamboo reinforced Manhole cover.

CONCLUSIONS

Light- duty bamboo reinforced Manhole cover is cost-effective and it will contribute towards sustainable development for a green alternative material to steel. Bamboo is not suitable for heavy-duty bamboo reinforced Manhole cover as its load carrying capacity is far behind the permissible limit as per IS: 2000(cl. 9.3).RCC and steel Manhole cover is being in danger of stealing as it has good scrap value in the black market which is a cause of many accidents and health problems. Bamboo reinforced manhole cover can be a solution to this problem as it has zero scrap value and the load-carrying capacity of the bamboo reinforced Manhole cover is nearby the Permissible limit as per IS: 2000(cl. 9.3). However, the further experimental investigation should be conducted to enhance the load-carrying capacity of the bamboo reinforced Manhole cover.

FUTURE SCOPE

The strength of bamboo reinforced light-duty Manhole cover can be enhanced further with the addition of fibers in concrete. Also, the use of Ferro-cement mesh with bamboo reinforcement may show improvement in the strength of the bamboo reinforced light duty manhole cover.

REFERENCES

[1] ASTM C478. Specification for precast reinforced concrete manhole sections. Philadelphia, USA: ASTM Standards Source; 2000.

[2] Ahmad, S., Raza, A. and Gupta, H. (2014) 'Mechanical Properties of Bamboo Fibre Reinforced 2.Concrete', 2nd International Conference on Research in Science, Engineering and Technology (ICRSET), Dubai.

[3] BS 497. Specification for manhole covers, road gully gratings and frames for drainage purposes. London, UK: British Standards Source; 2006.

[4] BS 7903. Guide to selection and use of gully tops and manhole covers for installation within the highway. London, UK: British Standards Source; 1997

[5] Indian Standard 12592 (2002) 'Precast Concrete Manhole Cover and Frame', first revision.,

[6] Indian Standard 456 (2000) 'Plain and Reinforced Concrete - Code of Practice', fourth revision.

[7] Khan, I. (2014) 'Bamboo Sticks as a Substitute of Steel Reinforcement in Slab', International Journal of Engineering and Management Research, 4(2), pp. 123-126.

[8] Muhtar (2020) 'Precast Bridges of Bamboo Reinforced Concrete in Disadvantaged Village Areas in Indonesia', Applied sciences 10(20):7158

[9] Ogunbiyi, A. et al. (2015) 'Comparative Analysis of the Tensile Strength of Bamboo and Reinforcement Steel Bars as Structural Member in Building Construction', International Journal of Scientific & Technology Research, 4(11), pp. 47-52.

[10] OÑA Mónica, SOLTANZADEH Fatemeh, DE SOUSA Christoph and BARROS Joaquim A. O.,(2015) "Exploring The Use of Hpfrc And Gfrp Grids for The Production of Manhole Covers", Fibre Concrete September 10-11, 2015, Prague, Czech Republic

[11] Patil, S. and Mutkekar, S. (2014) 'Bamboo as a Cost Effective Building Material for Rural Construction', Journal of Civil Engineering and Environmental Technology, 1(6), pp. 35-40.

[12] Salah El Haggar and Lama El Hatow (2009) 'Reinforcement of thermoplastic rejects in the production of manhole covers', Journal of Cleaner Production,Elsevier 17 (2009) 440–446

[13] "The World of Street Crime Has an Increasingly Popular Target: Manhole Covers". The New York Times Ed. May 4th (2012), page A22

[14] Weifeng, Z., Jing, Z. and Guobin, B. (2012) 'Application Technology of Bamboo Reinforced Concrete in Building', Applied Mechanics and Materials, 195-196, pp 297-302.

CITATIONS

- Dr. B.K.Sarkar, 202111051548, MAHATMA EDUCATION SOCIETY'S, PILLAI HOC COLLEGE OF ENGINEERING AND TECHNOLOGY, "Home Based Digital FM Transmission using Machine Learning and IoT based System", 2021, Indian Patent office.
- Dr. Lata Menon, Dr. Archana Bhagwat, Dr. Vishakha Telgote, 202121050103," MAHATMA EDUCATION SOCIETY'S PILLAI HOC COLLEGE OF ARTS, SCIENCE AND COMMERCE RASAYANI", Isolation and Characterization of Bioactive Unique Complex Compounds from Stembark Advanced Extract of Uapaca Pilosahutch, 2021, Indian Patent office.
- Dr. Lata Menon, Dr. Jayanta Kumar Behera, Dr. Sapana Chilate,202141051543, MAHATMA EDUCATION SOCIETY'S PILLAI HOC COLLEGE OF ARTS, SCIENCE AND COMMERCE RASAYANI TALUKA KHALAPUR DISTRICT RAIGAD MAHARASHTRA 410207, INDIA., " Power Generation from Sea Wave and Create Renewable Energy" 2021, Indian Patent office.
- Dr. B.K.Sarkar, 20211050099, MAHATMA EDUCATION SOCIETY'S, PILLAI HOC COLLEGE OF ENGINEERING AND TECHNOLOGY, "Online Technical Learning and Advanced Training Management System" 2021, Indian Patent office.
- Mr. Rajendra Bhimrao Deshmukh; 202121050101, MAHATMA EDUCATION SOCIETY'S, PILLAI COLLEGE OF EDUCATION AND RESEARCH, " Intelligent Biometric Sanitizer with Academy Tools" 2021, Indian Patent office.
- Dr. Mansi Subhedar (Associate Professor), Mr. Jayesh Rane (Assistant Professor),202121052899,MAHATMA EDUCATION SOCIETY'S; PILLAI HOC COLLEGE OF ENGINEERING &TECHNOLOGY;" Black Box for Two Wheelers" 2021, Indian Patent office.
- Dr. Priam Pillai (Associate Professor), Amar Arun Jadhav,(Assistant Professor)202121052898, MAHATMA EDUCATION SOCIETY'S, PILLAI COLLEGE OF ENGINEERING, PANVEL, "Antislip and Quick Release Grippers to Handle Soft and Delicate Materials" 2021, Indian Patent office.
- Munawira Kotyad (Associate Professor), Yvette Lee, (Social Worker and Counsellor), 202121052144, MAHATMA EDUCATION SOCIETY'S, PILLAI HOC COLLEGE OF ENGINEERING AND TECHNOLOGY, "Indian Child Specific Game Based

Psychotherapeutic (therapy for post-traumatic stress disorder" 2021, Indian Patent office.

- Sonam Prasad Bhopatrao, Prof (Dr.) Shilpa Pankaj Kewate, 202121052897, MAHATMA EDUCATION SOCIETY'S, PILLAI HOC COLLEGE OF ENGINEERING AND TECHNOLOGY, " Sustainable Sandwich Composite Circular Manhole Cover Designed for Pedestrian Networks" 2021, Indian Patent office.
- Dr. Shabab Ahmed Rizvi (Assistant Professor), Dr. Gajanan Wader (Principal), 202121052893, MAHATMA EDUCATION SOCIETY'S, Pillai College of Arts Commerce & Science, Dr K M Vasudevan PILLAI Campus, Sector 16, New Panvel, "E-O Indicator: Digital Engine oil indicator shows accurate conditions of engine oil" 2021, Indian Patent office.
- Nisha Subhash Yadav, 202121052895, MAHATMA EDUCATION SOCIETY'S, Pillai College of Arts Commerce & Science, Dr K M Vasudevan PILLAI Campus, Sector 16, New Panvel, NAVI Mumbai -410206, MH, India." Wireless Electric Current Transmitter and Receiver for Electronic Devices." 2021, Indian Patent office.
- Shardul Buva, Dr. Gajanan Pandit Wader, 202121053764,, MAHATMA EDUCATION SOCIETY'S Pillai College of Arts Commerce & Science, Dr K M Vasudevan PILLAI Campus, Sector 16, New Panvel, NAVI Mumbai -410206, MH, India, "Vehicle Spare Parts Identification System", 2021, Indian Patent office.
- Rohit Devappa Gawade, Prof(Dr.) Shilpa Pankaj Kewate, 202121054010, MAHATMA EDUCATION SOCIETY'S, PILLAI HOC COLLEGE OF ENGINEERING AND TECHNOLOGY, "Manufacturing of lightweight sandwich bricks using a unique combination of Bamboo Fiber and E-waste Powder" 2021, Indian Patent office.
- Dr. Sundeep Harish Deulkar, Dr. Arun Pillai, Dr. Malavika P. Sharma, 202121053765, MAHATMA EDUCATION SOCIETY'S, PILLAI COLLEGE OF ENGINEERING, " Use of Kinaesthetic Game Based Learning for Modelling the Hypothetical Energy Levels of A LASER", 2021, Indian Patent office.

CHAPTER - 04

Experimental Investigations on Bamboo Reinforced Manhole Cover as a Sustainable Approach

Pranav Parab[1], Dr. Shilpa Kewate[2]

[1]PG Student, Department of Civil Engineering, Pillai HOC College of Engineering and Technology, Raigad 410207, Maharashtra, India

[2]Associate Professor, Department of Civil Engineering, Pillai HOC College of Engineeringand Technology, Raigad 410207, Maharashtra, India

ABSTRACT

Eco-friendly and natural materials are given more importance for supporting sustainable development. The non-renewable materials such as steel is being replaced with bamboo which has a potential of getting similar results as steel reinforcement in concrete according to many studies. The process of selection, seasoning, treatment on bamboo will be carried out before for improving its properties and performance as reinforcement in concrete This research investigates the use of bamboo as reinforcement in the production of reinforced concrete manhole cover as alternative replacement to steel reinforcement. A main issue with manhole covers is stealing of cast iron manhole cover which may result in causing accidents due to hole left on road. The purpose of this study was to investigate the load-carrying capacity and the behavior of bamboo reinforced concrete manhole cover in order to apply this design in construction practice. Testing, grade, shape and size will be considered as per IS 12592:2002. This research work makes an approach to develop a new material which would contribute to sustainable development and also green alternative material to steel. This material will no longer become an issue in terms of theft and would contribute towards sustainable development.

Keywords: *Bamboo reinforcement; Green alternative; Manhole cover; Load test*

INTRODUCTION

Present Scenario

Due to vast development in whole world there is emission of harmful gases in environment from various industries which imbalances the ecosystem and a cause of adverse condition such as uneven rainfall, draughts, global warming, earthquakes, tsunami etc. This research has an approach to reduce carbon footprint in construction industries from steel and cement manufacturing process. For manufacturing of 1 ton of steel there is emission of approximately 2 ton of carbon dioxide and for 1 ton of cement there is approximately 1 ton of carbon dioxide in environment [1]. As due to increment of carbon footprint in environment the earth's temperature is rising year by year if this situation is continued then it will cause an unfavorable environment for human as well as other living organisms existing on earth and sustainability of life will be an issue. So there is a need to have an alternative and environmental friendly materials in construction which won't harm the environment and there will be sustainable development.

LITERATURE REVIEW

Bamboo is one such material which has a vast potential of replacing steel reinforcement in concrete. As it has good tensile as well as compressive strength which can be used as a green alternative to steel reinforcement. Using bamboo as reinforcement in concrete the strength and behavior of structure will be similar to steel reinforced concrete structure [2]. But before using bamboo as reinforcement it should be properly selected, seasoned, treated for achieving best results, compatible with concrete and to overcome some drawback [3]. The reinforcement ratio of bamboo in concrete is more than that of steel as it has less tensile strength than steel. Bamboo as reinforcement can be used for light weight RCC structures.

BAMBOO

Bamboo belongs to the sub-family of the *Bambusoideae* of the grass family *Poaceae*. It is a composite material of long and parallel fiber cellulose fibres embedded in a ligneous matrix [4]. It is a fast growing woody grass which can grow fully in 3-5 years with less water requirement and can sustain mostly in any type of soil and weather conditions if suitable species of bamboo is taken for cultivation according to the regional atmosphere. There are more than 1000 of bamboo species available in whole world.

Table 1: Advantages and disadvantages of bamboo

Advantages of bamboo	Disadvantages of bamboo
Low-cost	Tensile strength less than steel
Light weight compared to steel.	High water absorption due to which it swell
Bamboo can prevent pollution by absorbing large amounts of nitrogen from waste water.	Prone to termite attack and rotting
Reducing the amount of carbon dioxide in the air	Non uniform structure
Bio degradable	Low bond strength than steel with concrete
Available every where	

MANHOLE COVER

Manhole cover is a removable lid on the opening of manhole. The purpose of this is to avoid accidental falling anyone or anything into manhole cover. And also the manhole cover allows the authorized person for repair and maintenance work

Types of Manhole Cover

- Fibre reinforced plastic (FRP)
- Glass reinforced plastic (GRP)
- Reinforced cement concrete (RCC)
- Cast iron
- Steel fiber reinforced concrete (SFRC)
- Ductile iron

Shape of Manhole Ccover

- Circular
- Rectangular
- Square

Grade of Manhole Cover

Table 2: Grades of manhole cover. [5]

Grade	Grade Designation (Load capacity)	Uses purpose
Light Duty	LD-2.5 (25KN)	residential and institutional Complexes / areas with pedestrian, inspection chambers
Medium Duty	MD-10 (100KN)	Service lanes/roads, on pavements used under medium duty vehicular traffic, car parking areas
Heavy Duty	HD-20 (200KN)	institutional/commercial areas/ carriageways/city trunk roads /bus terminals with heavy duty vehicular traffic of wheel load between 50 to 100KN and parking areas and manhole chambers for road pavement
Extra Heavy Duty	EHD-35 (350KN)	carriageways in commercial/ industrial/port areas/ warehouses/ godowns where frequent loading and unloading of trucks/trailers, slow to fast moving vehicular traffic for wheel loads up to 115 KN

PRESENT STUDY

In this research there is investigation of bamboo reinforced manhole cover, bamboo reinforcement ratio is taken as 3% area of cross section which is test for load test of 25KN on 14 days from casting and also there is comparison of bamboo reinforced manhole cover with steel reinforced manhole cover for strength, weight and cost parameters. Bamboo reinforced manhole can also solve the issue of stealing of cast iron manhole cover which may result in causing accidents and health issue due to open hole left on road. And can also contribute towards sustainable development.

BAMBOO PREPARATION FOR REINFORCEMENT IN MANHOLE COVER

According to the literature survey bamboo cannot be used directly as reinforcement in concrete thus bamboo has to be properly selected, seasoned and treated before using. Also the flowchart showing the methodology of preparation process of manhole cover is shown in fig. 1

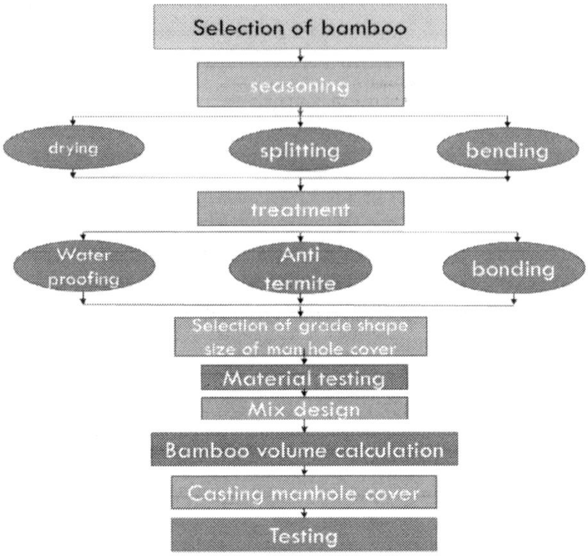

Figure 1: Flowchart of methodology

Selection

For reinforcement the bamboo selected should show brown color that means the bamboo is minimum of 3 years old. Longest and largest diameter of bamboo should be preferred [3]

Figure 2: Selection of largest diameter bamboo

Seasoning

Seasoning is a process that will make bamboo compatible for using as reinforcement in concrete

- Drying: selected bamboo should be well dried for 3-4 weeks because due to moisture the tensile strength of bamboo reduces [3]
- Splitting: bamboo cannot be used completely instead it should be split into strips of width size ranging from 18 to 25 mm.
- Bending: bamboo can bend with application of heat or pressure on it. (if required)

Figure 3: Splitting of bamboo in strips

Treatment

Waterproofing (Low water absorption): bamboo should be coated with coal tar for low water absorption as bamboo has the capacity to absorb upto 50% of water of its weight from concrete

- Anti-termite: coating or soaking in lethal solution to resist termite attack
- Good bonding in concrete: there should be sprinkling of sand on the coated bamboo with coal tar and dried for 24-48 hours [6]

Figure 4: Treatment on strips of bamboo with coal tar

Selection of Grade, Shape and Size of Manhole Cover

The grade, shape and size of manhole cover is selected as per IS code 12592:2002. In this study the grade selected is Light duty 2.5 which has a resisting capacity 25 KN load on it without any development of cracks on it, which is used for residential and institutional Complexes / areas with pedestrian but occasional light motor vehicle traffic, inspection chambers. Shape casted is square shape and size is 510mm x 510mm x 50mm in dimension

Material testing (if required): Raw material testing to ensure the quality of raw material used is as per the recommendation standard to achieve best results

Mix Design: For bamboo reinforced the water cement ratio must be as low as possible because bamboo can absorb water form concrete and the tensile strength of bamboo will reduce [7]

Bamboo volume calculation (ratio of bamboo reinforcement): For bamboo as reinforcement in concrete the ratio varies from 2.5- 4% of area of cross section. In this study the bamboo as reinforcement is taken as 3% of area of cross section in manhole cover

Casting of manhole cover: For casting process, the mesh of bamboo strips tied with binding wire of the required size is placed inside the mould (formwork) and placing of concrete with required amount of tamping or vibration is done

Testing on manhole cover: Load Test on manhole cover is done in UTM machine in this test application of target load i.e. 25 KN is considered with use of 300mm diameter circular block (plate) placed on manhole cover at its geometrical center and load application is done as per IS code 12592:2002 recommendation [5].

Figure 5: Test setup for testing of manhole cover in UTM

IS code 12592:2002 recommendation for making of manhole cover

Grade of concrete used must be 30 N/mm^2 or greater

Minimum cement content should be 360kg/m^3

Maximum water cement ratio of 0.45 can be adopted

Proper curing for 28 days should be adopted

Clear cover of minimum 15 mm is recommended [5]

Preparation of manhole cover

Materials

Cement

The Ordinary Portland cement of 53 grade

Fine aggregate:

Locally available sand conforming to grading zone 2 Sand passing through IS 4.75mm sieve

Coarse aggregate

Locally available crushed blue granite stones conforming to graded aggregate of nominal size 10mm

Water

Casting and curing of specimens is with the potable water that is available in the premises or on site.

Concrete Admixture

For good workability, water reduction, waterproofing, high performance

Mould (formwork)

As per selected shape and size of manhole cover

Binding wire

16gauge wire for tying reinforcement with each other

Coat tar

Coal tar liquid for bamboo waterproofing

Bamboo

Locally available Assam bamboo

Steel

FE-500 grade steel

Making of Manhole Cover

Total 4 manhole cover sample are casted of size 510mm x 510mm x 50mm and the grade of manhole cover targeted is LD-2.5 (25KN load resist capacity) which can be used for footpath, inspection chamber etc. The grade of concrete used is M30 grade for all manhole cover. 15mm clear cover is provided in all manhole cover. 3 sample of bamboo reinforced concrete manhole cover with 3% area of cross section as the reinforcement ratio and 1 sample of steel reinforced concrete with bar size of 8mm FE-500 grade steel are casted. Bamboo used is locally available Assam bamboo showing brown color with diameter of 50 mm and height of 4.8m long. Bamboo splitting into strips bamboo coated with coal tar with sand sprinkled, drying it for 48 hours and then used. Binding wires is used for tying bamboo as well as steel in mesh form. To lift manhole cover lifting forks of steel are provided for all manhole cover. For bamboo reinforced manhole cover only 1 lifting fork is allocated and for steel reinforced manhole cover 2 lifting fork are allocated for distinguishing purpose.

a **b**

Figure 6: (a) bamboo reinforcement before casting; (b) 3 sample of bamboo and 1 sample of steel reinforced concrete manhole cover.

RESULTS AND DISCUSSION

Weight Analysis

The weight of bamboo reinforced concrete manhole cover is differed drastically, also it is observed that the weight of a sample of bamboo reinforced manhole cover is more than steel reinforced concrete manhole cover. This may be due to non-uniform structure of bamboo or may be due to high water absorption capacity of bamboo, it may have absorbed the water from concrete during the process of curing. So proper drying and waterproofing treatment is necessary.

Table 3: Weights of manhole covers on 14th day.

510 x 510 x 50mm Manhole cover	Weight (Kg.)
Bamboo sample 1	30
Bamboo sample 2	30.35
Bamboo sample 3	29.5
Steel sample	30.15

Cost Analysis

From the cost analysis of manhole cover it is observed that the concrete required is same in bamboo as well as in steel reinforced manhole cover but the variation in cost is due to the reinforcement. The steel reinforcement cost is 3 times the cost of bamboo reinforcement that also includes its treatment cost. And the overall cost of bamboo reinforced manhole cover is approximately 40% less than steel reinforced concrete manhole

Table 4: Cost analysis of manhole cover.

510x510x50mm Manhole cover	Quantity	Price (Rs.) Steel manhole cover	Price (Rs.) Bamboo manhole cover
Concrete M30 Rs.5000/cum	0.013 cum	Rs.65	Rs.65
Reinforcement	2.37Kg steel for Rs.50000/ton Less than 1/3 bamboo used Rs.100 per Bamboo	Rs.118	Rs.30
Treatment	Coal tar	nil	Rs.10
Total cost of manhole cover		**Rs.183**	**Rs.105**

Results of Load Test on Manhole Cover

The load test on manhole cover is carried out to determine the load carrying capacity of manhole cover, as per the grade selected the manhole cover should resist 25 KN load without any cracking on the surface. Bamboo reinforced manhole cover sample has taken a peak load of 23 KN which is very closer to the targeted load and steel reinforced concrete manhole cover take the targeted load but there was visible crack observed on surface of cover.

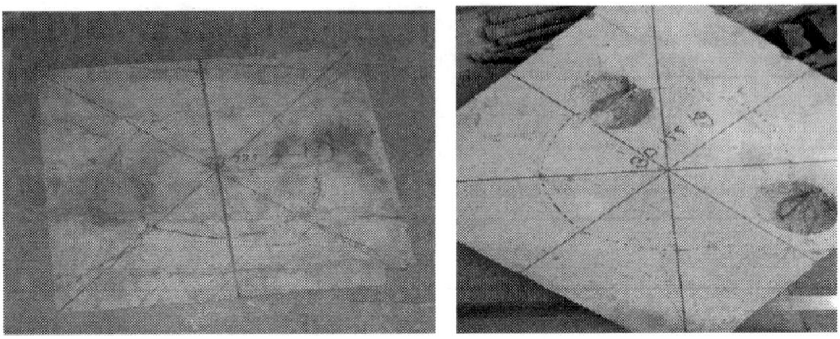

a b

Figure 7: (a) Bamboo reinforced concrete manhole cover failure pattern; (b) Steel reinforced concrete manhole cover crack on surface.

This testing is done on 14th day of casting to examine the strength at earlier stage. Final testing should be done by 28th day of casting.

Table 5: Load test results on manhole cover on 14th day.

510x510x50mm Manhole cover	Maximum Load taken (KN)
Bamboo sample 1	19
Bamboo sample 2	23
Bamboo sample 3	18
Steel sample	25

Graph plotted for weight of manhole cover vs load taken

The graph plotted presents weight of cover vs load taken by cover results in fig. 8

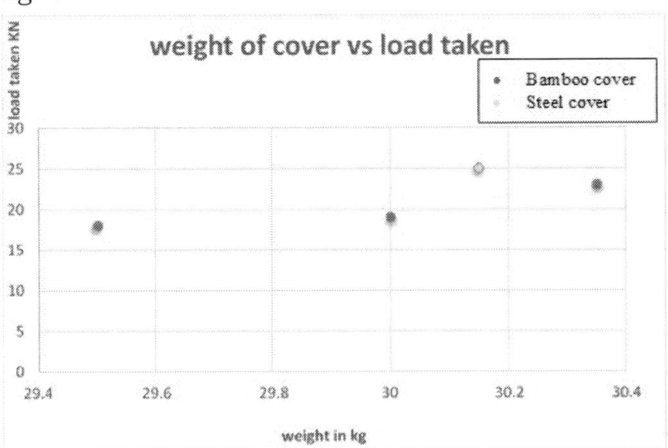

Figure 8: Weight of manhole cover vs load taken graph

FUTURE RESEARCH

Bamboo reinforced concrete manhole cover with 3% and 4% of area of cross section targeting for LD-2.5 (25KN) and MD-10 (100KN) grade by testing the bamboo reinforced concrete manhole cover on 28th day

Also many other structure can be targeted such as bamboo reinforced boundary wall, bamboo reinforced benches, low cost housing project etc. using bamboo reinforcement for sustainable development

CONCLUSIONS

1. Bamboo has very high water absorption capacity so proper water repellent coating is to be done on bamboo before using it as reinforcement
2. Bamboo reinforcement can reduce the cost of structure by approximately 40 % or more as compared to steel reinforcement
3. Bamboo reinforcement concrete manhole cover has shown a comparative good strength which may be improve till 28th day
4. Bamboo reinforcement should be given more attention as it can be successful alternative to steel reinforcement
5. Using bamboo reinforcement there can be reduction in carbon footprint from environment if used instead of steel and can also lower the effect of global harming
6. Bamboo reinforcement technology can surely contribute towards sustainable development

ACKNOWLEDGEMENTS

The author would like to thank his all the staff members of Pillai HOC College of Engineering, my family, friends for always supporting me and guiding me to solve all the issues in my project also special Thanks to structwel consultancy and there all staff members for allowing me to test my manhole cover and for providing further guidance for future project.

REFERENCES

[1] Suresh Bhalla, Supratic Gupta, Puttagunta Sudhakarand Rupali Suresh. Bamboo as green alternative to concrete and steel for modern structures. In: Journal of Environmental Research and Development Vol. 3 No. 2

[2] Ajinkya Kaware, Prof. U.R.Awari, Prof. M.R.Wakchaure. Analysis of Bamboo Reinforced Concrete Column. In: International Journal of Innovative Research in Science, Engineering and Technology, India, 2013, Vol. 2, Issue 6

[3] Anurag Nayak, Arehant S Bajaj, Abhishek Jain, Apoorv Khandelwal. Replacement of Steel by Bamboo Reinforcement. In: IOSR Journal of Mechanical and Civil Engineering, India, Jul - Aug. 2013, Volume 8, Issue 1 PP 50-61

[4] Khosrow Ghavami. Bamboo as reinforcement in structural concrete elements. In: Cement & Concrete Composites 27, Brazil, 2005. p. 637–649

[5] IS code 12592 (2002): Precast Concrete Manhole Cover and Frame.

[6] Dr. Shakeel Ahmad, Altamash Raza, and Hina Gupta. Mechanical Properties of Bamboo Fibre Reinforced Concrete. In: 2nd International Conference on Research in Science, Engineering and Technology (ICRSET'2014), Dubai, 2014

[7] Akash Priyadarshee1, Anil Kumar Chhotu2, Vikas Kumar. Use of Bamboo in Low Volume Rigid Pavement as Reinforced Material: A Review. In: Journal of Civil Engineering and Environmental Technology. 2014, Volume 1, Number 3

[8] Zhao Weifeng, Zhou Jing, Bu Guobin. Application Technology of Bamboo Reinforced Concrete in Building. In: Applied Mechanics and Materials. Xiangtan University, China, 2012 Vols. 195-196 p. 297-302

[9] Harish Sakaray, N.V. Vamsi Krishna Togati and I.V. Ramana Reddy. Investigation on Properties of Bamboo as Reinforcing Material in Concrete. In: International Journal of Engineering Research and Applications, India, 2012, Vol. 2, Issue 1, p.077-083

[10] I.K. Khan. Bamboo Sticks as a Substitute of Steel Reinforcement in Slab. In: International Journal of Engineering and Management Research, India, 2014, Volume-4, Issue-2, p. 123-126

[11] Jigar K. Sevalia, Nirav B. Siddhpura, Chetan S. Agrawal, Deep B. Shah, Jai V. Kapadia. Study on Bamboo as Reinforcement in Cement Concrete. In: International Journal of Engineering Research and Applications, India, 2013, Vol. 3, Issue 2, p.1181-1190

[12] Maruthupandian G., Saravanan.R, Suresh Kumar.S, Sivakumar.B.G. A Study on Bamboo Reinforced Concrete Slabs. In: Journal of Chemical and Pharmaceutical Sciences, India, 2016, Volume 9 Issue 2

[13] Ogunbiyi, Moses A., Olawale , Simon O ., Tudjegbe , Oke E ., Akinola , S . R. Comparative Analysis of the Tensile Strength of Bamboo And Reinforcement Steel Bars As Structural Member In Building Construction, In: International Journal of Scientific & Technology Research, 2015, Volume 4, Issue 11

[14] Ari Wibowo, Indradi Wijatmiko, and Christin Remayanti Nainggolan, Structural behavior of lightweight bamboo reinforced concrete slab with EPS infill panel, In: Green Construction and Engineering Education for Sustainable Future AIP Conf., Brawijaya University, Indonesia, 2017

[15] Dinesh Bhonde, P. B. Nagarnaik, D. K. Parbat, U. P. Waghe. Experimental Investigation of Bamboo Reinforced Concrete Slab. In: American Journal of Engineering Research, 2014, Volume-03 Issue-01 p-128-131

[16] Masakazu Terai & Koichi Minami. Research and Development on Bamboo Reinforced Concrete Structure. In: 15WCEE LISBOA. Fukuyama University, Japan, 2012

[17] Salah El Haggar, Lama El Hatow. Reinforcement of thermoplastic rejects in the production of manhole covers. In: Journal of Cleaner Production, American University, Egypt, 2009

[18] S. Karthik, P. Ram Mohan Rao, P.O. Awoyera. Strength properties of bamboo and steel reinforced concrete containing manufactured sand and mineral admixtures. In: Journal of King Saud University - Engineering Sciences, VIT University, India, 2016

CITATIONS

- GAURAV GYANENDRA MISHRA, PRAKASHDEV INDRADEO SINGH, DR. SHILPA PANKAJ KEWATE, PANKAJ KISHOR MAHALE, SALONEE RANGNEKAR, SHWETANG VASANT MESHRAM, AKASH YADAV, SNEHA SATISH MUSALE, VEDANT SANDEEP SAWANT, DEVESH THAKER, 202121053767, MAHATMA EDUCATION SOCIETY'S, PILLAI HOC COLLEGE OF ENGINEERING AND TECHNOLOGY, " Steel fibre reinforced sustainable composite hollow brick", 2021, Indian Patent office.
- Dr. C K PRASHANT, REMYA MADAN GOPAL, SUPARNA DEEPAK, 202121054595, MAHATMA EDUCATION SOCIETY'S, PILLAI HOC COLLEGE OF ARTS, SCIENCE AND COMMERCE, "Novel Radar Absorbing Paint for Military Stealth Technology." 2021, Indian Patent office.
- Prajakta Harshad Kane, 202121054600, MAHATMA EDUCATION SOCIETY'S, Pillai College of Arts Commerce & Science," Intelligent Computer Texter" 2021, Indian Patent office.
- Deepika Sharma, 202121054603, MAHATMA EDUCATION SOCIETY'S, Pillai College of Arts, Commerce & Science (Autonomous), Dr K M Vasudevan, "distance between user mobile phone and fitness band more than 5 mts (or user set distance) then alarm." 2021, Indian Patent office.
- Dr. Kavita Kathare, Dr. Aarti Sukheja, 202121054599, MAHATMA EDUCATION SOCIETY'S, Pillai College of Arts, Commerce & Science (Autonomous," show of the part which is harmed and must be supplanted in Washing machines" 2021, Indian Patent office.
- Dr. Divya Padmanabhan, Jweshvari Tupe, 202121054602,Mahatma Education Society's, Pillai College of Engineering "Natural Fiber Reinforced Composites for Filtration of Water" 2021, Indian Patent office.
- 49: Dr. Reni Francis (Principal), Dharti Latke, Dr. Jaya Cherian, 202121054601, MES's Pillai College of Education and Research " Facial Emotion Recognition using Deep Learning and AI." 2021, Indian Patent office.
- 50: Shaikh Farah Abbas, Dr. Reni Francis, 202121054604,MES's Pillai College of Education and Research "Improved the Relations of an International Trading Using AI" 2021, Indian Patent office.

- Achala Bhor, Dr. Mary George Varghese, 202121054605, MES's Pillai College of Education and Research, " Improve Your Restaurant's Online Food Ordering Using Machine Learning" 2021, Indian Patent office.
- Mange Amar Vilasrao, Jayesh Narahari Patil, Sagar Shivaji Mhatre, Sumit S Dalvi, 202121054596,MAHATMA EDUCATION SOCIETY'S, PILLAI HOC COLLEGE OF ENGINEERING AND TECHNOLOGY," automatic Signal Adjustment using IoT based Notification System." 2021, Indian Patent office.
- Mange Amar Vilasrao, Sumit S Dalvi, Jayesh Narahari Patil, Sagar Shivaji Mhatre, 202121054598, MAHATMA EDUCATION SOCIETY'S, PILLAI HOC COLLEGE OF ENGINEERING AND TECHNOLOGY," Agriculture Water Level Controlling using IoT-Based Technology" 2021, Indian Patent office.
- Mange Amar Vilasrao, SHIVRAJ PATIL, SWATI MHASKAR, PRADNYA RANE, ARJUN SATISH DESHMUKH, ANIKET PRAKASH DUMBRE, TATYASO ANNASO GARANDE, 202121054865,MAHATMA EDUCATION SOCIETY'S, PILLAI HOC COLLEGE OF ENGINEERING AND TECHNOLOGY, "Streetlights Monitoring using Machine learning and IoT-Based",2021, Indian Patent office.
- MRS DARSHANA JADHAV, DR. SALLY ENOS, 202121054866, MES' PILLAI COLLEGE OF EDUCATION AND RESEARCH Plot No 1, Sector-8, Khanda Colony, New Panvel (W), Raigad District, Maharashtra 410206, India.," Method and Process to Detection the Exam Cheating and P/F using Machine Learning" 2021, Indian Patent office.
- Mrs. Radhika Dave, Dr Bhavna Dave, 202121054869, MES' PILLAI COLLEGE OF EDUCATION AND RESEARCH MES' PILLAI COLLEGE OF EDUCATION AND RESEARCH, Plot No 1, Sector-8, Khanda Colony, New Panvel, " MPC-Academy Development: Method and Process Teaching and learning in higher education Using Machine Learning "2021, Indian Patent office.

Chapter - 05

Comparative Study on Bamboo and Steel Reinforced Light Duty Manhole Cover

Pranav Parab[1] and Dr. Shilpa Kewate[2]

[1]PG Student, Department of Civil Engineering, HOC Pillai College of Engineering and Technology, Rasayani 410207, Maharashtra, India. Email: pranavparab15041995@gmail.com

[2]Associate Professor, Department of Civil Engineering, HOC Pillai College of Engineering and Technology, Rasayani 410207, Maharashtra, India. Email: shilpakewate@mes.ac.in

ABSTRACT

One of the major environmental problem faced by the construction industry is carbon emission from steel and cement manufacturing. Due to the increment of carbon footprint in the environment, the earth's temperature is rising drastically. If the situation continues humans, as well as other living organisms on earth, may become extinct due to unfavourable environmental conditions. This study focuses on the reduction of the carbon footprint from steel manufacturing by replacing steel reinforcement with bamboo which is an environmentally friendly and renewable material. The major drawback of mild steel and reinforced steel manhole cover is stealing due to high scrap value. An uncovered hole left on the road may result in endangering lives. This research paper investigated the use of bamboo reinforcement as a substitute for steel in the construction of light duty manhole. In this paper manual design calculation, software analysis using Finite Element Analysis (FEM) with ANSYS software, experimental testing for load resisting capacity as per IS 12592:2002 of bamboo reinforced manhole cover, and also its comparison is shown. This research work approaches to contribute toward sustainable development of green alternative material to steel.

Keywords: *Bamboo reinforcement; Carbon footprint; Steel reinforcement; Sustainable development.*

INTRODUCTION

General

As there is huge industrialization in the whole world, the release of dangerous gases in the environment from various manufacturing sectors has imbalanced the ecosystem and a cause for unfavorable conditions such as unpredictable rainfall, rise in sea level, global warming, acidification of oceans, earthquakes, etc. One of the major issues faced by the construction industry is carbon emission from steel and cement manufacturing. For manufacturing one ton of steel approximately two tons of carbon dioxide is emitted and for one ton of cement one ton of carbon dioxide is released into the environment (Bhalla et al, 2008, p. 363). Due to the increment of carbon footprint in the environment, the earth's temperature is rising drastically. If these circumstances remain the same then it may make an unsuitable habitat for humans as well as other creatures existing on earth and feasibility of life will be an issue.

Concrete has good compression strength but low tensile strength. To overcome this condition in concrete structures, steel is introduced as reinforcement in the concrete for providing the tensile strength required for stability of the structure. Some issues commonly faced with steel are increment in overall cost, use of the non-renewable resources, polluting environment due to industrial processing, a major contributor to the overall dead load of structures, etc. So there is a requirement to have a substitute and a non-polluting material in construction that won't damage the ecosystem and has identical properties as steel which can support sustainable development.

Steel reinforced manhole cover and mild steel manhole cover have a good resale value due to which these manhole covers are in danger of being stolen which may cause accidents due to open holes left on the road. This may harm the lives of the public and animals walking on roads. The open holes of a manhole cover may disturb the lives of residing people due to bad odor and the arising of various diseases. Figure 1 shows an accident caused due to missing manhole cover on the road.

Light duty grade of manhole cover is used in the pedestrian way and inspection chamber in residential and institutional Complexes/ areas. As per IS 12592:2002 the minimum load resisting capacity of light duty manhole cover should be 25KN. In this research work, light duty bamboo reinforced manhole cover is designed manually, analyzed for deflection with finite element analysis (FEM) in ANSYS software, compared with steel reinforced manhole cover for weight, cost, carbon footprint perspectives, and tested in UTM for its load resisting capacity.

Figure 1: Accident due to missing manhole cover (Deshmukh, 2014).

Bamboo

Bamboo is a part of the grass family of Poaceae which belongs to the sub-family of Bambusoideae. It is a combined element made up of lengthy and parallel cellulose fibers compressed in a ligneous matrix form (Ghavami, 2004, p. 639). There is an abundant number of species of bamboo found in the whole world. The carbon sink by 5-year-old bamboo every year is 18 to 23 tons per acre.

Advantages of Bamboo

Bamboo is a natural, renewable, and biodegradable material. Its tensile strength to specific weight ratio is high, also it possesses both tensile and compressive strength (Nayak et al, 2013, p. 50). The cost of bamboo is cheaper as its availability is abundant, in rural as well as in the urban region. It can prevent pollution by absorbing a large amount of nitrogen from wastewater (Priyadarshee, Chhotu, and Kumar, 2014, p. 14).

Disadvantages of Bamboo

The bamboo tensile strength is relatively very lesser than steel (Khan, 2014, p. 126). Its water absorption capacity is very high which affects the dimension and develops low bond strength with concrete (Bhonde et al, 2014, p. 128). The structure of bamboo is irregular thus the strength may vary drastically. As it is a woody grass it can be attacked by termite and decay (Ahmad, Raza and Gupta, 2014, p. 163).

MANHOLE COVER

A manhole cover is a detachable covering on the hole of a manhole. The main role of this is to prevent accidentally falling anyone or anything inside the manhole. Also, it permits the approved individual for repair and maintenance work. The shapes of the cover available are circular, square, and rectangular.

Types of Manhole Cover

1. Fiber-reinforced plastic (FRP)
2. Glass-reinforced plastic (GRP)
3. Reinforced cement concrete (RCC)
4. Cast iron
5. Steel fiber reinforced concrete (SFRC)
6. Ductile iron.

Grade of Manhole Cover

The various grades of manhole cover and use purpose as Shown in Table 1

Table 1: Grade of manhole cover (IS 12592:2002).

Grade	Grade Designation (Load resist capacity)	Use purpose
Light Duty	LD-2.5 (25KN)	Residential and institutional Complexes/areas with pedestrian, inspection chambers
Medium Duty	MD-10 (100KN)	Service lanes/roads, on pavements used under medium duty vehicular traffic, car parking areas
Heavy Duty	HD-20 (200KN)	Institutional/commercial areas/ carriageways/ city trunk roads /bus terminals with heavy-duty vehicular traffic of wheel load between 50 to 100KN and parking areas and manhole chambers for road pavement

Grade	Grade Designation (Load resist capacity)	Use purpose
Extra Heavy Duty	EHD-35 (350KN)	Carriageways in commercial/industrial/port areas/ warehouses/ godowns where frequent loading and unloading of trucks/trailers, slow to fast-moving vehicular traffic for wheel loads up to 115 KN

METHODOLOGY

The methods adopted to achieve the desired outcome are explained. With a thorough analysis of the literature, a suitable methodology is fragmented. Manual design calculation, ANSYS software analysis, comparative analysis, and experimental testing of bamboo and steel reinforced manhole cover is conducted.

Manual Design Calculation of Bamboo Reinforced Manhole Cover

For the design calculation purpose, the manhole cover is considered as simply supported slab. Figure 2 shows the dimension of manhole cover

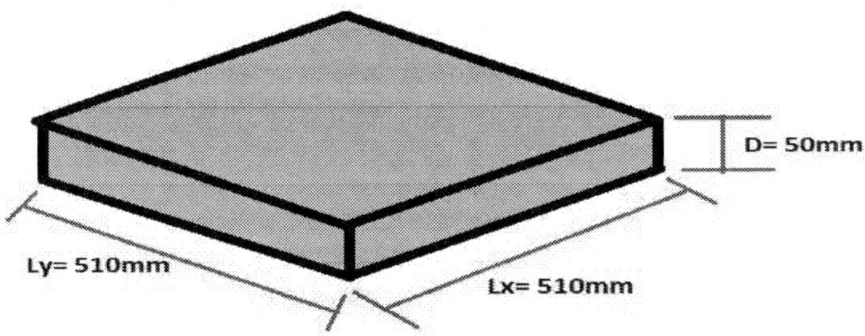

Figure 2: The dimension of manhole cover

Design calculation of light duty bamboo reinforced manhole cover dimension 510x510x50 mm

The grade of concrete taken is M30 and bamboo tensile strength is considered as 125 N/mm^2 (Sakaray et al, 2012, p. 84).

Considering bamboo bars as circular in cross-section as same as steel bars.

Step 1: Type of slab

Ly/Lx = 0.51/0.51 = 1 < 2 .....hence two-way slab.

Ly is the longest span and Lx is the shortest span of the slab.

Step 2: Depth of slab (IS 456:2000 clause 23.2.1)

d = span / BV * MF = 510 / 20*1.5 = 17mm

d is taken as 35mm for further calculation

D = 50mm

BV is the basic value, MF is modification factor, d is effective depth and D is total depth

Step 3: Effective span (IS 456:2000 clause 22.2)

L_{eff} = length of slab + effective depth (d) = 510 + 35 = 545 mm

Step 4: Load calculation

Dead load = density of RCC* volume = 25*0.035 = 0.875 KN/m

Live load = target load / length of slab = 25KN / 0.51m = 49.019 KN/m

Total Load = dead load + live load = 49.894 KN/m

Working load (W) =1.5* total load = 1.5 * 49.894 = 74.841 KN/m

Step 5: Ultimate moment (IS 456:2000 clause D-2.1)

αx=0.062,

αy=0.062 (IS 456:2000 Table 27)

Mx= $\alpha x * W * L_{eff}2$ = $0.062 * 74.841 * (0.545)^2$ = 1.378 KNm

My= $\alpha y * W * L_{eff}2$ = $0.062 * 74.841 * (0.545)^2$ = 1.378 KNm

Md (ultimate moment) = 1.378 KNm

αx and αy are moment coefficient

Step 6: Check for depth

Xu max = 0.46 d

Md = 0.36 * fck * Xu max * b (d- 0.42 Xu max)

$1.378 * 10^6$ = 0.36 * fck* 0.46 d *1000 *(d- 0.42*0.46 d)

d = 16.68 < 35 mm

fck is compressive strength of concrete

Step 7: Reinforcement calculation

Ast = (0.5*fck/fy) * (1-(in squareroot(1-((4.6*Md)/fck.bd^{2}))* bd

= (0.5*30/125)*(1-(insquareroot(1((4.6*1.378*10^{6})

/35*1000*35^{2}))*1000*35

Ast = 379.34 mm ^{2}

Ast is area of steel, fy is tensile strength of bar.

Step 8: Minimum reinforcement

0.12 % * bd = 0.12 / 100 * 1000 * 35 = 42 mm^{2}

Step 9: Spacing

Spacing = ((πd^{2}/4) / (Ast provided))* 1000 = ((3.14*10^{2}/4) / (379.34))*1000

= 207 mm but maximum spacing should be 300mm center to center

Hence use 5 bars of 10 mm diameter at 85mm center to center on x axis and y axis each d is the diameter of bar

Step 10: Check for shear

Vmax = W * Leff / 2 = 74.841*0.545 / 2 = 20.39 KN

max (tou max) = Vmax/ b *d = 20.39 * 10^{3} / 1000 * 35 = 0.58 N/ mm^{2}

□c (tou c)= ast / bd % = 0.89 N/ mm^{2}

(interpolation table 23 IS 456:2000)

$□_{permis}$ (tou permissible) = k * $□_{c}$ (tou c) = 1.3*0.89= 1.157 > $□_{max}$ (tou max=0.58) ok safe

Step 11: Check for deflection

Lx / d <= α β γ λ δ

Lx / d = 0.51 / 0.035 = 14.57

α = 20, β = 1, γ =1, λ =1, δ =1

fs= (0.58* fy* ast req)/ ast prov = (0.58*125* 379.34) / 393 = 69.98

γ= 1.6 (figure 4 IS 456:2000)

α β γ λ δ = 36 = 1.6*1*1*1*20

Lx / d <= α β γ λ δ hence 14.57 < = 36 ok safe

Similarly, for steel reinforced light duty manhole cover of size 510x510x50mm grade of concrete taken is M30 and FE 500 grade steel. 5 number of steel bars of 8 mm @ 85mm c/c on x axis and y axis each are used.

ANSYS Software Analysis

In this software, the manhole cover is reinforced with bamboo and steel as per the above design, and dimension is modeled. This model is analyzed for deflection and then compared with the deflection limit as per IS 456:2000 clause 23.2 (Yadav and Tambe, 2016, p. 1001). The steps involved in modeling and analysis in ANSYS software are displayed in the flowchart in Figure 3.

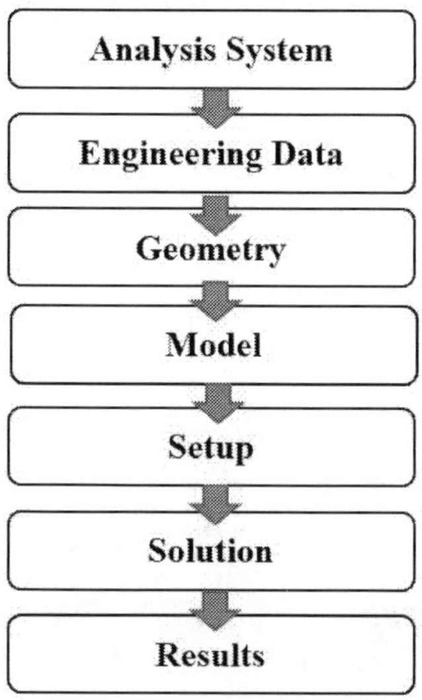

Figure 3: Flowchart of steps involved in ANSYS

Analysis System

In ANSYS software there are various analysis systems such as fluid flow, explicit dynamics, transent structural, etc. for the current analysis work Static structural system is used. In static structural system analysis of strain, stress, displacements in the geometry due to assigning of boundary condition and application of load is analyzed.

Engineering Data

In engineering data, we have to define material properties or select the materials which will be used for analysis in further steps. We can find a large number of material properties already defined in this software. Also, the software allows editing or redefining the properties of the material as per the requirement of the user. Table 2 shows the mechanical properties of steel, bamboo, and concrete used for analysis.

Table 2: Mechanical properties of steel, bamboo, and concrete (Sakaray et al, 2012, p. 84, Raju and Poluraju, 2019, p. 69)

Mechanical properties	Steel	Bamboo	Concrete
Density	7850 kg/m^3	693 kg/m^3	2400 kg/m^3
Young's modulus	2x10^{11} Pa	1.73x10^{10} Pa	3x10^{10} Pa
Poisson's ratio	0.3	0.384	0.18
Bulk modulus	1.6667x10^{11} Pa	2.4856x10^{10} Pa	1.5625x10^{10} Pa
Shear modulus	7.6923x10^{10} Pa	6.25x10^9 Pa	1.2712x10^{10} Pa
Tensile yield strength	5x10^8 Pa	1.25x10^8 Pa	-
Tensile ultimate strength	5.45x10^8 Pa	2.26x10^8 Pa	5x10^6 Pa
Compressive yield strength	2.5x10^8 Pa	-	3x10^7 Pa

Geometry

In this step, the user has to draw the geometrical model of the body on which the analysis has to be conducted. For this, the interface used is design modeler, which consists of various drawing tools to make the required geometry as a user requires. The dimension of the manhole cover considered for analysis is 510x510x50mm Figure 4 shows the geometry made of the manhole cover.

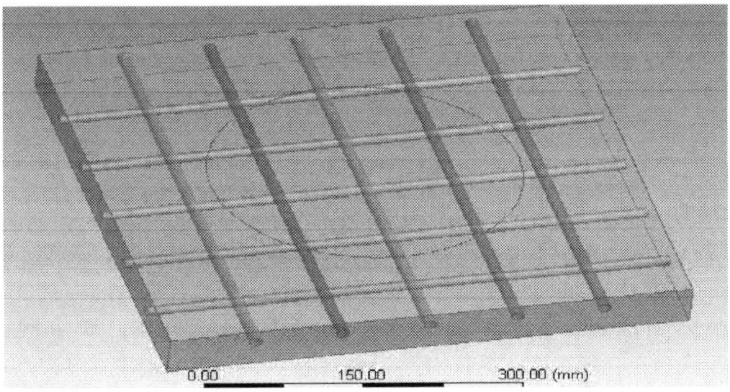

Figure 4: Manhole cover geometry in design modeler

Model

In model the user has to assign the properties to every solid body or geometry element that the user has defined in engineering data such as bamboo, concrete, or steel. Figure 5 shows the meshing of manhole cover and Figure 6 shows the meshing of reinforcement.

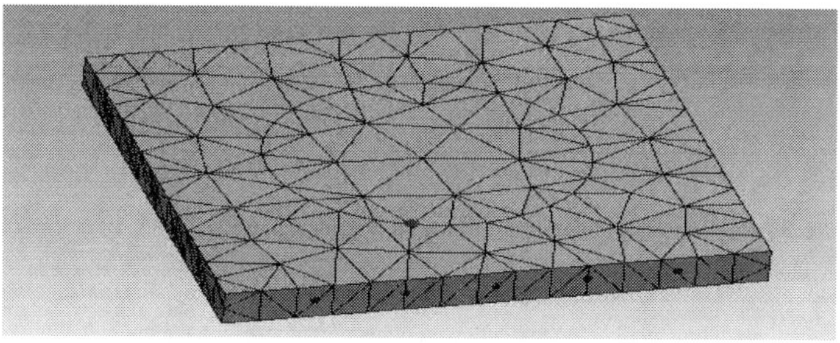

Figure 5: Meshing of manhole cover

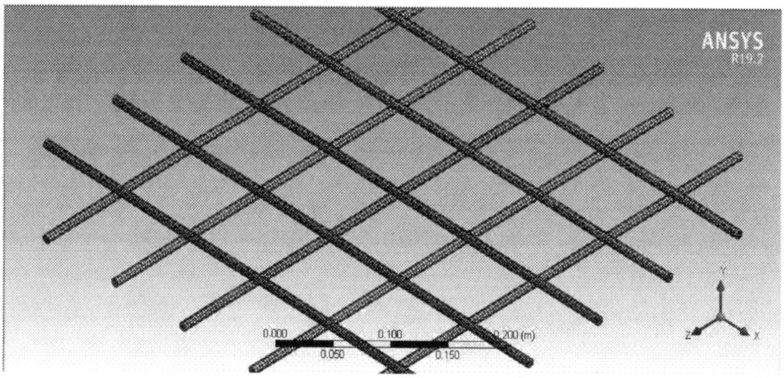

Figure 6: Meshing of reinforcement

Setup

In this step, the user defines the boundary condition to the model such as support, force, velocity, etc. for this analysis 25KN load is applied in the center of a manhole cover with 300 diameter area Figure 7 show supports, and Figure 8 shows the load applied on in the center with 300mm diameter area on the manhole cover.

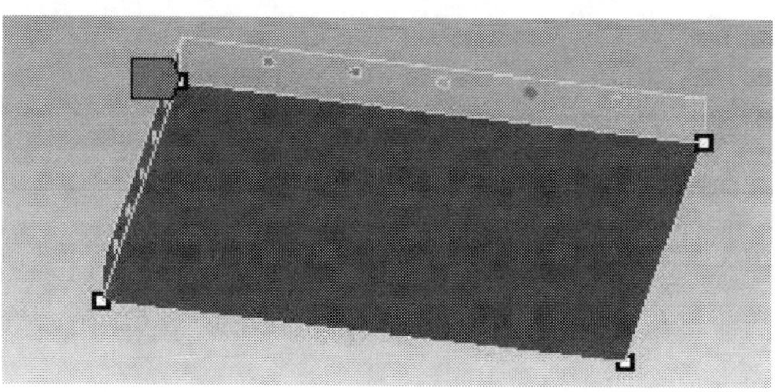

Figure 7: Support for manhole cover

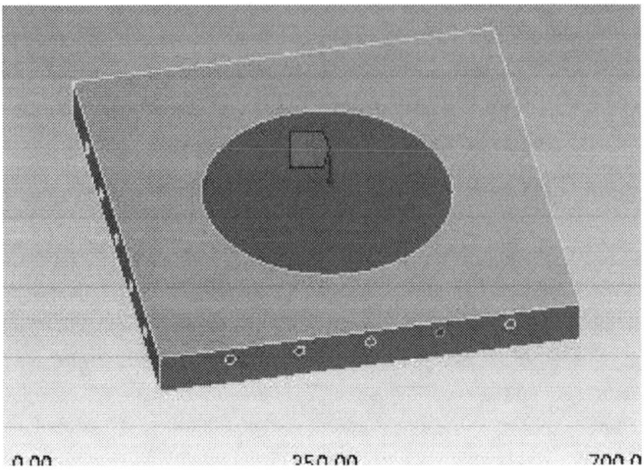

Figure 8: Load applied in the center of manhole cover

Solution

In this step, the user can define the desired requirement from the model. We can define stress, strain, total deformation (deflection), etc. of the model for getting its results.

Results

In this step, the user gets the answers for the desired requirements of his model.

Experimental Testing of Manhole Cover

The casting of the manhole cover

According to the literature, survey bamboo cannot be used directly as reinforcement in concrete thus bamboo has to be dried to reduce moisture content, then spitted vertically into strips form as per desired dimension (area). The cross-sectional area of single bamboo strips was taken as $78.5mm^2$. Strips are then coated with coal tar and sprinkled sand on them which are well dried for 48hrs (Brink and Rush, 1966). This makes the bamboo withstand water absorption and termites and sprinkling of sand makes a good bond with concrete. Bamboo mesh tied with binding wire and then cast with M30 grade concrete. The water-cement ratio taken should be minimum as possible as bamboo can absorb excess water from concrete (Brink and Rush 1966). 3 sample of bamboo and steel reinforced manhole cover is cast for testing purpose. Figure 9 shows bamboo strips treated with coal tar and Figure 10 shows a mesh of bamboo before casting of the manhole cover.

Figure 9: Bamboo strips treated with coal tar.

Figure 10: Bamboo mesh for manhole cover.

Testing on manhole cover (load resisting capacity of manhole cover)

Testing of the manhole cover is conducted in UTM as per the arrangement displayed in Figure 11 as referred to in IS 12592:2002. The load is applied on the manhole cover with a rate of 1 KN/s upto its target by keeping a circular plate or block of 300mm diameter on its geometrical center. The grade considered for this experiment is light duty which resists 25KN as the target load.

Figure 11: Arrangement in UTM for testing manhole cover.

In the experimental work, the two cases of manhole cover are considered case 1 steel reinforced and case 2 bamboo reinforced manhole cover. The comparison of weight, cost, carbon footprint, and load resisting capacity is analyzed Table 3 shows the details of the two cases of manhole cover.

Table 3: Details of case 1 and case 2.

Case number	Number of samples	Type	Reinforcement	Dimension (mm)
Case 1	3	Light Duty	Steel (5 bars of 8 mm on x and y axis each)	510x510x50
Case 2	3	Light Duty	Bamboo (5 strips of 10 mm on x and y axis each)	510x510x50

RESULTS

Deflection Analysis of Manhole Cover in ANSYS Software

The analytical work is done for target load and deflection (deformation) for the modeling of bamboo and steel reinforced manhole cover. These results are compared with the limits of deflection with IS 456:2000 clause 23.2 page number 37 i.e. span/ 350. Table 4 show the details and deflection results of manhole cover Figure 12 and 13 show the deflection in steel and bamboo manhole cover respectively.

Table 4: Details and deflection results of manhole cover

Case	Type	Dimension (mm)	reinforce-ment	Deflection (mm)	Deflection Limit (mm)
Case 1	Light duty	510x510x50	Steel	0.63497	510/350
Case 2	Light duty	510x510x50	Bamboo	0.64236	=1.457

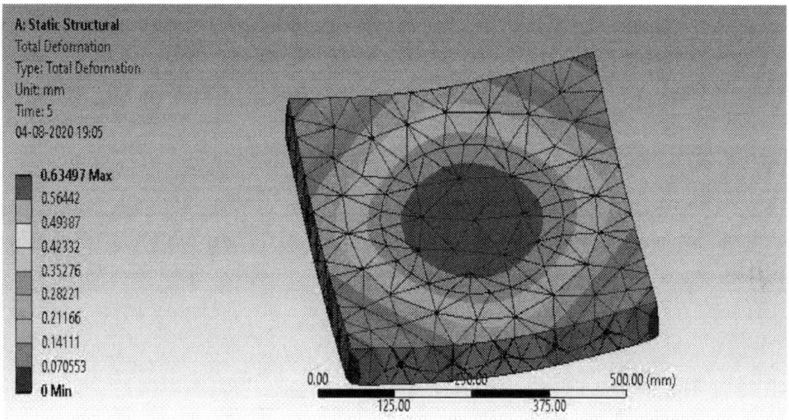

Figure 12: Case 1 deflection in steel manhole cover.

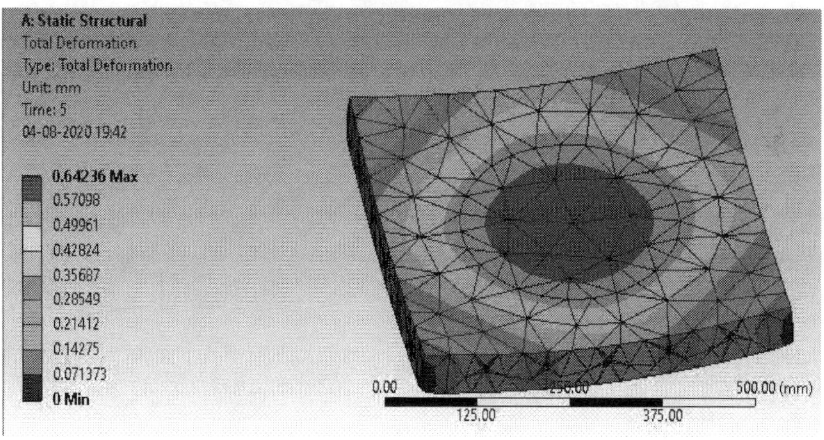

Figure 13: Case 2 deflection in bamboo manhole cover.

Experimental Analysis and Testing of Manhole Cover

Weight analysis

From the weight comparison of manhole cover, the weight of bamboo reinforced cover case 2 is observed to be lighter than steel reinforced manhole cover case 1. Table 5 shows the details of the weight analysis of manhole cover.

Table 5: Weight analysis of manhole cover.

Case number	Dimension (mm)	reinforce-ment	Weight (kg.)	Average weight (kg.)
Case 1: light duty steel reinforcement	510x510x50	Steel	30.15	30.3
	510x510x50	Steel	30.3	
	510x510x50	Steel	30.5	
Case 2: light duty bamboo reinforcement	510x510x50	Bamboo	29.5	29.95
	510x510x50	Bamboo	30	
	510x510x50	Bamboo	30.35	

Cost analysis

The cost of bamboo reinforced cover case 2 is less as compared to steel reinforced manhole cover case 1 from the cost analysis is observed.

Case 1: light duty steel reinforced manhole cover.

8 mm diameter 5 number of steel bars with spacing 85mm center to center in x and y direction each for casting manhole cover.

The length of one steel bar is 0.51m and 10 bars were used.

Cutting length of steel = 0.51* 10 = 5.1 m

Weight of steel in kg. = 5.1 *D^2/ 162 = 2.01 kg.

(D= diameter of steel bar)

Concrete quantity = 0.51x0.51x0.05= 0.013 cum

Table 6 shows cost analysis of case 1 size 510x510x50mm light duty steel reinforcement manhole cover.

Table 6: Cost analysis of case 1 steel manhole cover.

Case 1 light duty steel reinforcement	Concrete M30 (Rs. 5000/cum)	Reinforcement (kg.) (Rs. 50000/ ton Steel)	Total (Rs.)
Quantity	0.013	2.01	
Price Rs.	65	100.5	165.5

Case 2: light duty bamboo reinforced manhole cover

Bamboo reinforcement is assumed to be circular as steel reinforcement.

Area of single bamboo strip = 78.5 mm^2

10mm diameter 5 number of bars with spacing 85 mm center to center in x and y direction each for casting manhole cover.

Less than one-third of bamboo was used.

Concrete quantity = 0.51x0.51x0.05= 0.013 cum.

Table 7 show cost analysis of case 2 size 510x510x50mm light duty bamboo reinforcement manhole cover.

Table 7: Cost analysis of case 2 bamboo manhole cover.

Case 2 light duty 3% bamboo reinforcement	Concrete M30 (cum) (Rs. 5000/ cum)	Reinforcement (Rs. 100 per bamboo)	Treatment coal tar	Total (Rs.)
Quantity	0.013	Less than 1/3 bamboo used	Rs. 10	
Price (Rs.)	65	30	10	105

Carbon footprint analysis

The major contribution of carbon footprint is for steel and cement thus only these two are considered for analysis. The carbon footprint from 1 ton of cement is approximately 1 ton and from 1 ton of steel, approximately 2 tons of carbon is emitted (Bhalla et al, 2008, p. 363). The carbon footprint from case 2 bamboo manhole is less as compared to case 1 steel manhole cover Figure 14 shows the graph of the carbon footprint from case 1 and 2 manhole cover.

Case 1: light duty steel reinforced manhole cover:

Weight of cement used in case 1 = volume of manhole cover* kg. of cement per cum =0.013*380 = 4.94 kg of cement.

Carbon from concrete (cement) = 4.94 * 1 = 4.94 kg.

Weight of steel is taken from the above calculation of cost analysis of case 1.

Weight of steel = 2.01 kg.

Carbon from reinforcement (steel) = 2.01 * 2= 4.02 kg.

Table 8 show carbon footprint analysis of case 1 size 510x510x50mm light duty steel reinforcement manhole cover.

Table 8: Carbon footprint from case 1 steel manhole cover.

Case 1 light duty steel reinforcement	Concrete	Reinforcement	Total (kg.)
Carbon footprint (kg.)	4.94	4.02	8.96

Case 2: light duty bamboo reinforced manhole cover:

Cement calculation same as case 1= 4.94 kg of cement

Carbon footprint from concrete (cement) = 4.94 * 1 = 4.94 kg.

Carbon footprint from reinforcement (bamboo) = 0

Table 9 show carbon footprint analysis of case 2 size 510x510x50mm light duty bamboo reinforcement manhole cover.

Table 9: Carbon footprint from case 2 bamboo manhole cover.

Case 2 light duty bamboo reinforcement	Concrete	Reinforcement	Total (kg.)
Carbon footprint (kg.)	4.94	-	4.94

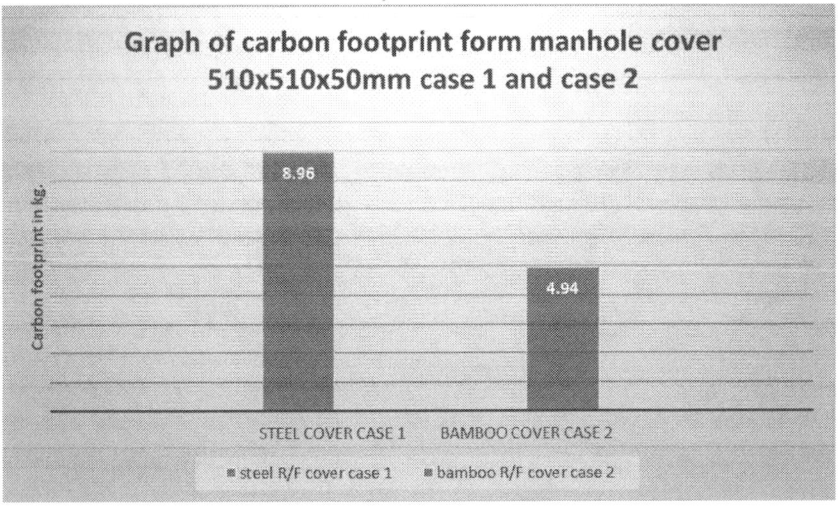

Figure 14: Graph of the carbon footprint from case 1 and 2 manhole cover

Load resisting capacity (strength) of manhole cover

The target load for light duty manhole cover case 1 and case 2 is 25KN. If the manhole cover resists the target load the load is not increased further. From load test results all three steel reinforced manhole cover sample case 1 has successfully resisted the target load. Whereas the maximum load resisted by the bamboo reinforced manhole cover sample is 23KN and the average load resisted is 20KN. However, the strength of bamboo reinforced manhole cover can be enhancing further by conducting more trials with adding fibers (steel, bamboo, polypropylene, glass fiber, etc.) in concrete, using Ferro-cement mesh, by increasing bamboo reinforcement ratio, etc. Table 10 shows the load test results and Figure 15 shows the cracks observed on the bamboo reinforced manhole cover after testing.

Table 10: Load test results (load resisting capacity) of manhole cover.

Case number	Dimension (mm)	reinforce-ment	Load resisted (KN)	Aver-age load (KN)	Remark
Case 1 light duty steel reinforced	510x510x50	Steel	25	25	As per IS 12592:2002 the minimum load resisting capacity should be 25KN
	510x510x50	Steel	25		
	510x510x50	Steel	25		
Case 2 light duty bamboo reinforced	510x510x50	Bamboo	18	20	
	510x510x50	Bamboo	19		
	510x510x50	Bamboo	23		

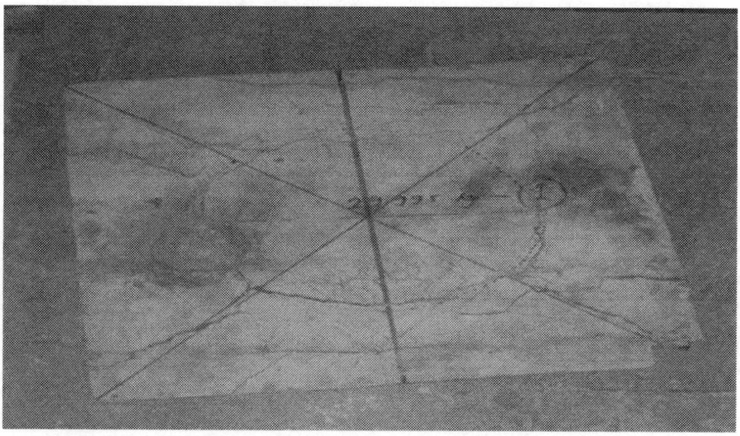

Figure 15: Crack on bamboo reinforced manhole cover after testing.

CONCLUSIONS

Following are the conclusions of this study:

- From the software analysis in ANSYS, steel and bamboo reinforced manhole cover both are found to be within the permissible limit of deflection as per IS 456:2000 clause 23.2.
- The weight of bamboo reinforced manhole case 2 is lighter as compared to steel reinforcement in manhole cover case 1.
- The total cost of a bamboo reinforced manhole cover is 36% cheaper as compared to a steel reinforced manhole cover.

- Carbon footprint can be reduced by 44% with the use of bamboo reinforcement in the manhole cover.
- The load resisting capacity of bamboo reinforcement manhole cover case 2 of light duty grade has resisted a maximum load of 23KN which is very close to the target load and the average load resisted is 20KN while steel reinforced manhole cover case 1 has resisted the target load of 25KN. Further trials and testing is required to make bamboo a successful eco-friendly alternative to steel in the reinforced manhole cover.
- The use of bamboo reinforcement can lower the consequences of global warming which can promote sustainable development.

FUTURE SCOPE

The strength of bamboo reinforced light duty manhole cover can be enhanced further with the addition of fibers (steel, bamboo, polypropylene, glass fiber etc.) in concrete or by increasing the area of bamboo etc.

Also, the use of Ferro-cement mesh with bamboo reinforcement may show improvement in the strength of the bamboo reinforced light duty manhole cover.

REFERENCES

[1] Bhalla, S. et al. (2008) 'Bamboo as Green Alternative to Concrete and Steel for Modern Structures', *Journal of Environmental Research and Development*, 3(2), pp. 362-370.

[2] Deshmukh, C. (2014) 'Biker Sustains Severe Injuries As Manhole Cover Gives Way', *Mid-day*, 22 November [online]. Available at: https://www.mid-day.com/articles/biker-sustains-severe-injuries-as-manhole-cover-gives-way/15780174

[3] Ghavami, K. (2005) 'Bamboo as reinforcement in structural concrete elements', *Cement & Concrete Composites*, 27 (2005), pp. 637–649.

[4] Indian Standard 12592 (2002) 'Precast Concrete Manhole Cover and Frame', first revision.

[5] Indian Standard 456 (2000) 'Plain and Reinforced Concrete - Code of Practice', fourth revision.

[6] Sakaray, H. et al. (2012) 'Investigation On Properties of Bamboo as Reinforcing Material in Concrete', *International Journal of Engineering Research and Applications*, 2(1), pp. 77-83.

[7] Bhonde, D. et al. (2014) 'Experimental Investigation of Bamboo Reinforced Concrete Slab', *American Journal of Engineering Research*, 3(1), pp. 128-131.

[8] Brink, F. and Rush, P. (1966) 'Bamboo Reinforced Concrete Construction', *U. S. Naval Civil Engineering Labaoratory Port Hueneme*, California.

[9] Yadav, P. and Tambe, H. (2016) 'Analytical study of solid flat slab and voided slab using ANSYS Workbench', *International Research Journal of Engineering and Technology,* 3(10), pp. 1000-1003.

[10] Khan, I. (2014) 'Bamboo Sticks as a Substitute of Steel Reinforcement in Slab', *International Journal of Engineering and Management Research,* 4(2), pp. 123-126.

[11] Nayak, A. et al. (2013) 'Replacement of Steel by Bamboo Reinforcement', *IOSR Journal of Mechanical and Civil Engineering,* 8(1), pp. 50-61.

[12] Priyadarshee, A., Chhotu, A. and Kumar, V. (2014) 'Use of Bamboo in Low Volume Rigid Pavement as Reinforced Material: A Review', *Journal of Civil Engineering and Environmental Technology,* 1(3), pp. 14-16.

[13] Dr. Ahmad, S., Raza, A. and Gupta, H. (2014) 'Mechanical Properties of Bamboo Fibre Reinforced Concrete', *2nd International Conference on Research in Science, Engineering and Technology (ICRSET),* Dubai. Available at: http://dx.doi.org/10.15242/IIE.E0314522

[14] Raju, B. and Poluraju P. (2019) 'Numerical Study of Bubble Deck Slab Using Ansys', *International Journal of Recent Technology and Engineering,* 7(6C2), pp. 68-72.

CITATIONS

- Mrs. Isha Dhody, Dr. Geeta S Thakur, 202121054870, MES' PILLAI COLLEGE OF EDUCATION AND RESEARCH, Plot No 1, Sector-8, Khanda Colony, New Panvel (W), Raigad District, Maharashtra 410206, India.," Students Education Monitoring using Machine Learning and IoT Based Technology." 2021, Indian Patent office.
- Mrs. Suparna Deepak, Dr. Navami Dayal, Dr. C. K. Prashant , 202121055060, MAHATMA EDUCATION SOCIETY'S, Pillai College of Arts, Commerce & Science (Autonomous), Dr. K. M. Vasudevan Pillai Campus, Plot No.10, Sector 16, New Panvel, Navi Mumbai, Maharashtra 410206, India.," Phytochemical properties and vitro antidiabetic capability of blend of 6 Indian restorative plants." 2021, Indian Patent office
- Dr. Navami Dayal, Suparna Deepak, 202121054868, MAHATMA EDUCATION SOCIETY'S, Pillai College of Arts, Commerce & Science (Autonomous), Dr. K. M. Vasudevan Pillai Campus, Plot No.10, Sector 16, New Panvel, Navi Mumbai, Maharashtra 410206, India., "Composite Micronutrient Nano fertilizer", 2021, Indian Patent office.
- Ms. Bindu Rajaguru, Ms. Meenakshi Johri, Mr. Gopakumar Pillai. 202121055061, MAHATMA EDUCATION SOCIETY'S, Pillai College of Arts, Commerce & Science (Autonomous), Dr. K. M. Vasudevan Pillai Campus, Plot No.10, Sector 16, New Panvel, Navi Mumbai, Maharashtra 410206, India.," Integrated Green Tea Infused with Jamun Seeds and Gum Arabic." 2021, Indian Patent office.
- Dr. B.K. Sarkar, US-17551376, MAHATMA EDUCATION SOCIETY'S, PILLAI HOC COLLEGE OF ENGINEERING &TECHNOLOGY RASAYANI, TALUKA PANVEL, DIST, NAVI MUMBAI, MAHARASHTRA 410207, INDIA., " A Method for COVID-19 Detection, Spread Prevention, Medical Assistance Using Machine Learning and Deep Learning, 2021, US- Patent Office.
- Dr. Reni Francis, G11237DE, MES's Pillai College of Education and Research Chembur. "Effectiveness of STEAM education in fostering Integrative learning through iLEARN Model." German Patent Office.

- Dr. Sanjay B. Munavalli, 202121056484, Pillai College of Arts, Commerce & Science (Autonomous)
- Dr K M Vasudevan PILLAI Campus, Sector 16, New Panvel, NAVI Mumbai -410206, MH, India, "Intelligent Digital Water Purifier (IDWP) with Filter Quality Indicator (FQI) and Water Consumption Indicator (WCI)" 2021, Indian Patent office.
- Dr. Abida Khan, 202121056482 Vasudevan PILLAI Campus "Department of B.Com. Accounting & Finance, Pillai College of Arts, Commerce & Science (Autonomous Dr K M, Sector 16, New Panvel, NAVI Mumbai -410206, MH, India, " Colouring Crayons/Pencils which kids can use to scribble only on surface of notebook/pages and not on wall/floor/furniture." 2021, Indian Patent office.
- Dr. Kiran Sagar Deshmukh, 202121056485, Department of B. Com, Pillai College of Arts Commerce and Science (Autonomous), Dr K M Vasudevan PILLAI Campus, Sector 16, New Panvel, NAVI Mumbai -410206, MH, India ," Intelligent Electric Vehicle Battery Supper Fast Stations" 2021, Indian Patent office.
- Dr. C K Prashant, Suparna Deepak, 202121056488," Department of Biotechnology, Pillai college of Arts, Commerce & Science (Autonomous), Dr K M Vasudevan PILLAI Campus, Sector 16, New Panvel, NAVI Mumbai -410206, MH, India, "Method and Process for Targeting Intra-Cellular Bacterial Infections Using Nanoparticles. "2021, Indian Patent office
- Ms. Remya Madan Gopal, 202121056486, MAHATMA EDUCATION SOCIETY'S Pillai HOC College of Arts, Science and Commerce, Pillai HOCL Educational Campus, Rasayani, Taluka Khalapur, Dist, Raigad, Maharashtra 410207, "Cancer Therapeutics Using Nanoparticles Based on Warburg Effect." 2021, Indian Patent office.
- MAHATMA EDUCATION SOCIETY'S ,202121056483, MAHATMA EDUCATION SOCIETY'S PILLAI HOC COLLEGE OF ENGINEERING AND TECHNOLOGY, "Experimental Studies on the Use of Waste Rubber Tyre Crumb and Bamboo in Tremix Concrete Flooring" 2021, Indian Patent office.
- Suprita Manohar Rao, Sadhana Shivkumar Thevar, 202121057477, Mahatma Education Society's Pillai College of Arts, Commerce & Science (Autonomous), Dr K M Vasudeva PILLAI Campus, Sector 16, New Panvel, NAVI Mumbai -410206, MH, India, "Intelligent Methods and Process to Rc- Cycle the Waste Plastic" 2021, Indian Patent office.

- Ms. Meenakshi Johri, Ms. Bindu Rajaguru, Mr. Gopakumar Pillai, 202121057473, MAHATMA EDUCATION SOCIETY'S PILLAI COLLEGE OF ARTS, COMMERCE & SCIENCE (AUTONOMOUS), " Nutraceutical Product Combinig Wheat Grass and Fenugreek" 2021, Indian Patent office.

CHAPTER - 06

Comparative Analysis on Bamboo Reinforced Manhole Cover as a Sustainable Approach

[1]Pranav Parab, [2]Dr. Shilpa Kewate

[1] P.G. Student, Department of Civil Engineering, Pillai HOC College of Engineering and Technology, Rasayani-410207, India

[2]Associate Professor, Pillai HOC College of Engineering and Technology, Rasayani-410207, India

ABSTRACT

Eco-friendly and natural materials are generally given more importance for supporting sustainable development. The non-renewable materials such as steel is being replaced with bamboo which has a potential of getting similar results as steel reinforcement in concrete according to many studies. The process of selection, seasoning, treatment on bamboo is carried out before, for improving its properties and performance as reinforcement in concrete This research paper investigates the use of bamboo as reinforcement in the production of reinforced concrete manhole cover as alternative replacement to steel reinforcement. A main issue with manhole covers is stealing of cast iron manhole cover which may result in causing accidents due to hole left on road. The purpose of this study is to comparisons bamboo reinforced concrete manhole with steel reinforced manhole cover on the basis of preliminary analysis. Testing, grade, shape and size is considered as per IS 12592:2002. This research paper makes an approach to develop a new material which would contribute to sustainable development and also a green alternative material to steel reinforcement.

Keywords: *Bamboo Reinforcement, Carbon Footprint, Green Alternative, Load Test, Manhole Cover.*

INTRODUCTION

Due to vast development in whole world there is emission of harmful gases in environment from various industries which imbalances the ecosystem and a cause of adverse condition such as uneven rainfall, draughts, global warming, earthquakes, tsunami etc. This research has an approach to reduce carbon footprint in construction industries from steel and cement manufacturing process. For manufacturing of 1 ton of steel there is emission of approximately 2 ton of carbon dioxide and for 1 ton of cement there is approximately 1 ton of carbon dioxide in environment (Suresh et al., 2008). Due to increment of carbon footprint in environment the earth's temperature is raising year by year. If this situation is continued, then it will cause an unfavorable environment for human as well as other living organisms existing on earth and sustainability of life will be an issue. So there is a need to have an alternative and environmental friendly material in construction which won't harm the environment and which can contribute towards sustainable development.

Bamboo is one such material which has a vast potential of replacing steel reinforcement in concrete. As has good tensile as well as compressive strength which can be used as a green alternative to steel reinforcement. Using bamboo as reinforcement in concrete the strength and behavior of structure will be similar to steel reinforced concrete structure (Ajinkya et al., 2013). But before using bamboo as reinforcement it should be properly selected, seasoned, treated for achieving best results, making compatible with concrete and to overcome some drawback (Anurag et al., 2013). The reinforcement ratio of bamboo in concrete is more than that of steel as it has less tensile strength than steel. Bamboo as reinforcement can be used for light weight RCC structures.

Bamboo

Bamboo belongs to the sub-family of the *Bambusoideae* of the grass family *Poaceae*. It is a composite material of long and parallel fiber cellulose fibers embedded in a ligneous matrix (Khosrow 2005). It is a fast growing woody grass which can grow fully in 3-5 years with less water requirement and can sustain mostly in any type of soil and weather conditions if suitable species of bamboo is taken for cultivation according to the regional atmosphere. There are more than 1000 of bamboo species available in whole world. Bamboo is used in construction majorly of scaffolding and roofing purpose. Now-a-days bamboo flooring and other decorative uses of bamboos are done in building system for achieving green rating for a particular project. Bamboo is one such material which can be an alternative material to wood as much attractive furniture can be made with use of bamboo. Bamboo can be used for treatment of rabies, Arthritis, urinal infection, mouth and gum sores, inflammations and irritability etc. Also research claim that there can be generation of bio-fuel with use of bamboo. Table 1 shows advantages and disadvantages of bamboo.

Table 1: Advantages and disadvantages of bamboo.

Advantages of bamboo	Disadvantages of bamboo
• Low-cost • Light weight compared to steel. • Bamboo can prevent pollution by absorbing large amounts of nitrogen from waste water. • Reducing the amount of carbon dioxide in the air • Bio degradable • Available every where	• Tensile strength less than steel • High water absorption due to which it swell • Prone to termite attack and rotting • Non uniform structure • Low bond strength than steel with concrete

MANHOLE COVER

Manhole cover is a removable lid on the opening of manhole. The purpose of this is to avoid accidental falling of anyone or anything into manhole. The manhole cover allows the authorized person for repair and maintenance work.

Types of Manhole cover

- Fiber reinforced plastic (FRP)
- Glass reinforced plastic (GRP)
- Reinforced cement concrete (RCC)
- Cast iron
- Steel fiber reinforced concrete (SFRC)
- Ductile iron.

Shape of Manhole Cover

- Circular
- Rectangular
- Square.

Grade of Manhole Cover

Grades of manhole cover are shown in Table 2.

Table 2: Grades of manhole cover. (IS 12592:2002).

Grade	Grade Designation (Load capacity)	Uses/ purpose
Light Duty	LD-2.5 (25KN)	residential and institutional Complexes / areas with pedestrian, inspection chambers
Medium Duty	MD-10 (100KN)	Service lanes/roads, on pavements used under medium duty vehicular traffic, car parking areas
Heavy Duty	HD-20 (200KN)	institutional/commercial areas/ carriageways/city trunk roads /bus terminals with heavy duty vehicular traffic of wheel load between 50 to 100KN and parking areas and manhole chambers for road pavement
Extra Heavy Duty	EHD-35 (350KN)	carriageways in commercial/ industrial/port areas/ warehouses/ godowns where frequent loading and unloading of trucks/trailers, slow to fast moving vehicular traffic for wheel loads up to 115 KN

METHODOLOGY

In this research paper preliminary analysis on bamboo reinforced manhole cover is carried out. Bamboo reinforced manhole can also solve the issue of stealing of cast iron manhole cover which may result in causing accidents and health issue due to open hole left on road. And can also contribute towards sustainable development. Bamboo reinforcement ratio is considered as 3% of area of cross section. Grade considered is light duty 2.5 with dimension 510x510x50mm. Comparison of bamboo reinforced manhole cover with steel reinforced manhole cover for cost and carbon footprint is done and water absorption test is done on bamboo.

According to the literature survey bamboo cannot be used directly as reinforcement in concrete thus bamboo has to be properly selected, seasoned and treated before using. Also the flowchart showing the methodology of preparation process of manhole cover is shown in figure 1.

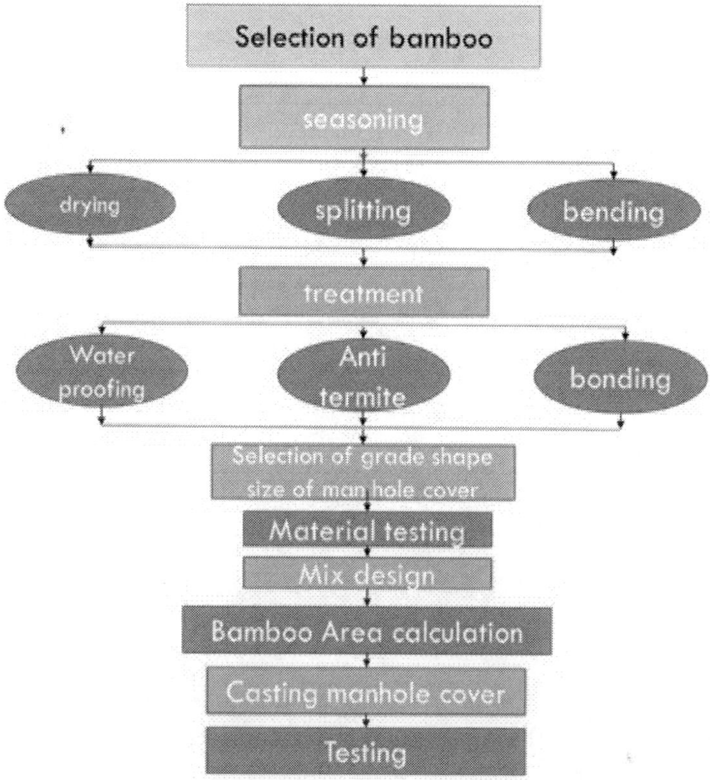

Figure 1: Flowchart of Methodology

Selection

For reinforcement the bamboo selected should show brown color that means the bamboo is minimum of 3 years old. Longest and largest diameter of bamboo should be preferred (Anurag et al., 2013).

Figure 2: Selection of largest diameter bamboo

Seasoning

Seasoning is a process that will make bamboo compatible for using as reinforcement in concrete.

- **Drying:** selected bamboo should be well dried for 3-4 weeks because due to moisture the tensile strength of bamboo reduces (Anurag et al., 2013).
- **Splitting:** bamboo cannot be used completely instead it should be split into strips of width size ranging from 18 to 25 mm.
- **Bending:** bamboo can bend with application of heat or pressure on it. (if required)

Figure 3: Splitting of bamboo in strips

Treatment

- **Waterproofing (Low water absorption):** bamboo should be coated with coal tar for low water absorption as bamboo has the capacity to absorb upto 50% of water of its weight.
- **Anti-termite:** coating or soaking in lethal solution to resist termite attack.
- **Good bonding in concrete**: sprinkling of sand on the coated bamboo with coal tar and dried for 24-48 hours (Shakeel et al., 2014).

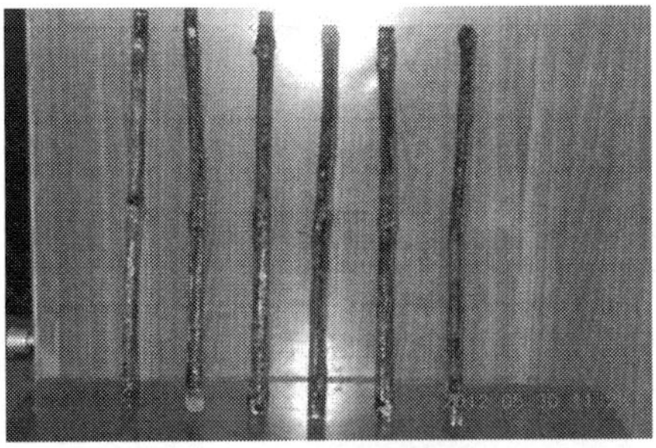

Figure 4: Treatment on strips of bamboo with coal tar (Shakeel et al., 2014).

Selection of Grade, Shape and Size of Manhole Cover

The grade, shape and size of manhole cover is selected as per IS code 12592:2002. In this study the grade considered is Light duty 2.5 which has a resisting capacity 25 KN load on it without any development of cracks on it, which is used for residential and institutional Complexes / areas with pedestrian but occasional light motor vehicle traffic, inspection chambers. Shape casted is square shape and size is 510mm x 510mm x 50mm in dimension.

Material Testing

Raw material testing to ensure the quality of raw material used is as per the recommendation standard to achieve best results.

Mix Design

Grade of concrete considered is M30. For bamboo reinforced the water cement ratio must be as low as possible because bamboo can absorb water form concrete and the tensile strength of bamboo will reduce (Akash et al., 2014). Table 3 shows M30 mix design.

Table 3: Mix design of M30 grade concrete

Sl. No.	Materials	kg/m3
1	Cement	380
2	Water	160
3	Fine Aggregate	711
4	Coarse Aggregate	1283
	20 mm coarse Aggregate	924
	10 mm coarse Aggregate	359
5	Admixture	1.90
6	Water Cement Ratio	0.42

Bamboo Area Calculation (Ratio of Bamboo Reinforcement)

For bamboo as reinforcement in concrete the ratio varies from 2.5-4% of area of cross section. In this study the bamboo as reinforcement is taken as 3% of area of cross section of manhole cover.

Casting of Manhole Cover

For casting process, the mesh of bamboo strips tied with binding wire of the required size is placed inside the mould (formwork) and placing of concrete with required amount of tamping or vibration is recommended.

Testing on Manhole Cover

Load Test on manhole cover is done in UTM machine in this test application of target load i.e. 25 KN is applied use of 300mm diameter circular block (plate) placed on manhole cover at its geometrical center and load application will be made as per IS code 12592:2002 recommendation. Figure 5 shows the setup of UTM for testing manhole cover.

Figure 5: Test setup for testing of manhole cover in UTM

RESULTS AND DISCUSSION

Water Absorption Test on Bamboo

3 sample of bamboo strips are taken which were dried natural in sunlight were weighed and immersed in water for 24 hours. The average water absorption by bamboo is 44% by its weight. Table 3 shows the water absorption test results.

Table 3: Water absorption by bamboo sample

	Dried weight grams	**Immersed weight**	**Water absorption %**
Sample 1	95	136	43
Sample 2	102	149	46
Sample 3	107	155	44
		average	44

Cost Analysis

From the cost analysis of manhole cover it is observed that the concrete required is same in bamboo as well as in steel reinforced manhole cover but the variation in cost is due to the reinforcement. The steel reinforcement cost is 3 times the cost of bamboo reinforcement that also includes its treatment cost. The overall cost of bamboo reinforced manhole cover is approximately 40% less than steel reinforced concrete manhole. Table 5 shows the cost analysis of manhole cover.

Table 5: Cost analysis of manhole cover.

510x510x50mm Manhole cover	**Quantity**	**Price (Rs.) Steel manhole cover**	**Price (Rs.) Bamboo manhole cover**
Concrete M30 Rs.5000/cum	0.013 cum	Rs.65	Rs.65
Reinforcement	2.37Kg steel for Rs.50000/ton, Less than 1/3 bamboo will be used Rs.100 per Bamboo	Rs.118	Rs.30
Treatment	Coal tar	nil	Rs.10
Total cost of manhole cover		Rs.183	Rs.105

Carbon Footprint Analysis

For manufacturing of 1 ton cement there is 1 ton carbon emitted and for 1 ton steel manufacturing there is 2 ton carbon emitted in environment (Suresh et al., 2008). Bamboo reinforced manhole cover has controlled 4.74 kg that is 49% less carbon footprint than steel reinforced manhole cover. Table 7 shows carbon footprint analysis.

Table 7: Carbon footprint analysis

	Bamboo manhole cover (kg.)	**Steel manhole cover (kg.)**
Carbon from cement	4.94	4.94
Carbon from steel reinforcement	-	4.74
Total	4.97	9.68

FUTURE SCOPE

Research can be conducted on bamboo reinforced boundary wall, bamboo reinforced benches, low cost housing project using bamboo reinforcement etc. for Sustainable development.

CONCLUSIONS

- Bamboo has very high water absorption capacity so proper water repellent coating has to be done on bamboo before using it as reinforcement.
- Bamboo reinforcement can reduce the cost of structure by approximately 40 % or more as compared to steel reinforcement.
- Bamboo reinforcement should be given more attention as it can be successful alternative to steel reinforcement.
- Using bamboo reinforcement there can be reduce carbon footprint from environment and can also lower the effect of global harming.
- Bamboo reinforcement technology can surely contribute towards sustainable development.

ACKNOWLEDGEMENTS

The author would like to thanks all the staff members of Pillai HOC College of Engineering, my family, friends for always supporting me and guiding me to solve all the issues in my project.

REFERENCES

[1] Ajinkya Kaware, Prof. U.R.Awari, Prof. M.R.Wakchaure. 2013. Analysis of Bamboo Reinforced Concrete Column. International Journal of Innovative Research in Science, Engineering and Technology, India, Vol. 2, Issue 6.

[2] Akash Priyadarshee1, Anil Kumar Chhotu, Vikas Kumar.2014. Use of Bamboo in Low Volume Rigid Pavement as Reinforced Material: A Review. Journal of Civil Engineering and Environmental Technology. Volume 1, Number 3.

[3] Anurag Nayak, Arehant S Bajaj, Abhishek Jain, Apoorv Khandelwal. 2013. Replacement of Steel by Bamboo Reinforcement. IOSR Journal of Mechanical and Civil Engineering, India,Volume 8, Issue 1 PP 50-61.

[4] Ari Wibowo, Indradi Wijatmiko, and Christin Remayanti Nainggolan. 2017. Structural behavior of lightweight bamboo reinforced concrete slab with EPS infill panel. Green Construction and Engineering Education for Sustainable Future AIP Conf., Brawijaya University, Indonesia.

[5] Dinesh Bhonde, P. B. Nagarnaik, D. K. Parbat, U. P. Waghe. 2014. Experimental Investigation of Bamboo Reinforced Concrete Slab. American Journal of Engineering Research, Volume-03 Issue-01 p-128-131.

[6] Dr. Shakeel Ahmad, Altamash Raza, and Hina Gupta. 2014. Mechanical Properties of Bamboo Fibre Reinforced Concrete. 2nd International Confer-

ence on Research in Science, Engineering and Technology (ICRSET 2014), Dubai.

[7] Harish Sakaray, N.V. Vamsi Krishna Togati and I.V. Ramana Reddy. 2012. Investigation on Properties of Bamboo as Reinforcing Material in Concrete. International Journal of Engineering Research and Applications, India. Vol. 2, Issue 1, p.077-083.

[8] I.K. Khan. 2014. Bamboo Sticks as a Substitute of Steel Reinforcement in Slab. International Journal of Engineering and Management Research, India, Volume-4, Issue-2, p. 123-126.

[9] IS code 12592 (2002): Precast Concrete Manhole Cover and Frame.

[10] Jigar K. Sevalia, Nirav B. Siddhpura, Chetan S. Agrawal, Deep B. Shah, Jai V. Kapadia. 2013. Study on Bamboo as Reinforcement in Cement Concrete. International Journal of Engineering Research and Applications, India, Vol. 3, Issue 2, p.1181-1190.

[11] Khosrow Ghavami. 2005. Bamboo as reinforcement in structural concrete elements. Cement & Concrete Composites 27, Brazil. p. 637–649.

[12] Maruthupandian G., Saravanan.R, Suresh Kumar.S, Sivakumar.B.G. A Study on Bamboo Reinforced Concrete Slabs. Journal of Chemical and Pharmaceutical Sciences, India, 2016, Volume 9 Issue 2.

[13] Masakazu Terai & Koichi Minami. 2012. Research and Development on Bamboo Reinforced Concrete Structure. 15WCEE LISBOA. Fukuyama University, Japan.

[14] Ogunbiyi, Moses A., Olawale, Simon O ., Tudjegbe, Oke E ., Akinola, S . R. 2015. Comparative Analysis of the Tensile Strength of Bamboo And Reinforcement Steel Bars As Structural Member In Building Construction. International Journal of Scientific & Technology Research, Volume 4, Issue 11.

[15] S. Karthik, P. Ram Mohan Rao, P.O. Awoyera. 2006. Strength properties of bamboo and steel reinforced concrete containing manufactured sand and mineral admixtures. Journal of King Saud University - Engineering Sciences, VIT University, India.

[16] Salah El Haggar, Lama El Hatow. 2009 Reinforcement of thermoplastic rejects in the production of manhole covers. Journal of Cleaner Production, American University, Egypt.

[17] Suresh Bhalla, Supratic Gupta, Puttagunta Sudhakarand Rupali Suresh. 2008 Bamboo as green alternative to concrete and steel for modern structures. Journal of Environmental Research and Development Vol. 3 No. 2.

[18] Zhao Weifeng, Zhou Jing, Bu Guobin. 2012. Application Technology of Bamboo Reinforced Concrete in Building. Applied Mechanics and Materials. Xiangtan University, China, Vols. 195-196 p. 297-302.

CITATIONS

- Prof. Suhas Lawand, Prof. Rupali Nikhare, Dr. Prashant Nitnaware, Prof. Deepti L., Prof. Gayatri Hegde, 202121057471, MAHATMA EDUCATION SOCIETY'S, PILLAI COLLEGE OF ENGINEERING, "IoT based System for Women and Children Safety", 2021, Indian Patent office.
- Dinesh G Tiwari, 202121057474, MAHATMA EDUCATION SOCIETY'S, PILLAI COLLEGE OF ENGINEERING, Dr K M Vasudevan PILLAI Campus, Sector 16, New Panvel, NAVI Mumbai -410206, MH, India, "Solar Power Emulator" 2021, Indian Patent office
- Dr. Sharda Sharma,G11244DE, MAHATMA EDUCATION SOCIETY'S, Dr. Pillai Global Academy " Mathoscope" German Govt
- Dr. Daphne Pillai, Dr. Priam Pillai, Franav Pillai, Deepika Sharma, Dr. Kavita Kathare, MAHATMA EDUCATION SOCIETY'S, Pillai College of Arts Commerce & Science(Autonomous), "Skill Credit System" German Govt.
- Dr. R. H. Khade, Professor, Electronics & Computer Science, Dr. Swati R. Khade, Yash R. Khade, Gaurav R. Khade 202121058522, MAHATMA EDUCATION SOCIETY'S, PILLAI COLLEGE OF ENGINEERING, Dr K M Vasudevan PILLAI Campus, Sector 16, New Panvel, NAVI Mumbai -410206, MH, India, "additional collars of same or different colour/cloth/print along with shirt." 2021, Indian Patent office.
- Mr. Shubham Thakur, Mr. Vishnu Khadasane, Mr. Pratik Mhatre, Ms. Snehal Patil, Ms. Sneha Jadhav, Mr. Amar Mange 202121059583, MAHATMA EDUCATION SOCIETY'S, PILLAI HOC COLLEGE OF ENGINEERING &TECHNOLOGY," Automatic Hand Sanitizer Control Board." 2021, Indian Patent office
- Nikam Ravindra Rajaram, Nikam Sonali Ravindra, Nikam Vaishnavi Ravindra, Nikam Vishwanath Anand 202121059584, MAHATMA EDUCATION SOCIETY'S, PILLAI COLLEGE OF ENGINEERING, Dr K M Vasudevan PILLAI Campus, Sector 16, New Panvel, NAVI Mumbai -410206, MH, India," Highly reflecting coatings on front windshield of the car" 2021, Indian Patent office
- Mrs. Vaishali Kumbhar, Mr Amey Marathe, 202121061527, MAHATMA EDUCATION SOCIETY'S PILLAI COLLEGE OF ARTS, COMMERCE & SCIENCE (AUTONOMOUS)," Vehicle Speed Automatically Lock if objects Detects under Defined Distance." 2021, Indian Patent office

- Dr. G R Patil, Dr. K D Singh, Raju Narwade, Karthik Nagarajan, 202121061528, MAHATMA EDUCATION SOCIETY'S PILLAI HOC COLLEGE OF ENGINEERING &TECHNOLOGY RASAYANI, TALUKA PANVEL," Slope bottom tuned liquid damper for improving performance of flat bottom tuned liquid damper." 2021, Indian Patent office
- Dr. G R Patil, Raju Narwade, Karthik Nagarajan, Prof.(Dr.) B.K.Sarkar, 202121062299, MAHATMA EDUCATION SOCIETY'S PILLAI HOC COLLEGE OF ENGINEERING &TECHNOLOGY RASAYANI," New scrubber for cleaning and washing of stainless tiffin box and utensils/pot used in cooker." 2021, Indian Patent office

CHAPTER - 07

Manhole Cover with Bamboo as Sustainable Reinforcement in Light Duty Grade

[1]Pranav Umesh Parab, [2]Prof.(Dr.) Shilpa Pankaj Kewate

[1]Mahatma Education Society's
[2]Pillai Hoc College of Engineering & Technology Rasayani, Taluka Panvel, Dist, Navi Mumbai, Maharashtra 410207, India.

ABSTRACT

Our invention gives importance to eco-friendly and natural materials for supporting sustainable development. The non-renewable materials such as steel are being replaced with bamboo which has the potential of getting similar results as steel reinforcement in the concrete according to many studies. The process of selection, seasoning, and treatment of bamboo is carried out before to improve its properties and performance as reinforcement in concrete this research investigates the use of bamboo as reinforcement in the production of reinforced concrete manhole cover as an alternative replacement to steel reinforcement. The main issue with manhole covers is stealing of cast iron and steel reinforced manhole cover which may result in causing accidents due to open holes left on road. The purpose of this study was to investigate the load resisting capacity of bamboo reinforced concrete manhole cover in order to apply this design in construction practice. Testing, grade, and shape are considered as per IS |12592:2002|. This research work makes an approach to develop a new material that would contribute to sustainable development and also green alternative material to steel. This material will no longer become an issue in terms of theft and would contribute towards sustainable development.

FIELD OF THE INVENTION

Our Invention is related to Manhole Cover with Bamboo as Sustainable Reinforcement in Light Duty Grade.

BACKGROUND OF THE INVENTION

As there is enormous industrialization globally, the release of dangerous gases in the environment from various manufacturing sectors has imbalanced the ecosystem. And a cause for unfavorable conditions such as unpredictable rainfall, rise in sea level, global warming, acidification of oceans, earthquakes, etc. and also the crucial problem encountered by the construction sector is carbon emission from steel and cement manufacturing. The construction sector contributes up to 11% of carbon dioxide emissions globally.

For manufacturing 1 ton of steel, there is an emission of approximately 2 tons of carbon dioxide, and for manufacturing 1 ton of cement, there is an emission of approximately 1 ton of carbon dioxide in the environment. Due to the increase of carbon footprint in the environment, the earth's temperature rises year by year. If this situation continues, it will cause an unfavorable environment for humans and other living organisms, and the sustainability of life will be an issue.

Concrete has good compressive strength but low tensile strength. To overcome this condition in concrete structures, steel is introduced as reinforcement in concrete for providing the tensile strength required for the stability of the structure. However, some issues commonly faced with steel are increment in overall cost, use of the non-renewable resources for its manufacturing, polluting environment due to industrial processing, a significant contributor for the overall dead load of structures, etc. So there is a need to have an alternative and environmentally friendly material in construction that won't harm the environment, similar to steel, which can contribute to sustainable development.

Steel reinforced manhole cover, and mild steel manhole cover has a good resale value due to which these manhole covers are in danger of being stolen, which may cause accidents due to open holes left on the road. Furthermore, it may harm the life of the public and animals walking on roads. Again, the open holes of the manhole may disturb the lives of residing people due to foul odors and arising to various diseases

BAMBOO

Bamboo is a part of the grass family of Poaceae, which belongs to the sub-family of Bambusoideae. It is a combined element made up of lengthy and parallel cellulose fibers compressed in a ligneous matrix form. There is an abundant number of species of bamboo found in the whole world. The carbon sink by 5-year-old bamboo every year is 18 to 23 tons per acre.

Advantages of Bamboo

Bamboo is a natural, renewable, and biodegradable material. Its tensile strength to specific weight ratio is high. Also, it possesses both tensile and compressive strength. Therefore, the cost of bamboo is cheaper as its availability is abundant in rural and urban regions. In addition, it can prevent pollution by absorbing a large amount of nitrogen from wastewater.

Disadvantages of Bamboo

The bamboo tensile strength is relatively much lesser than steel (Khan, 2014, p. 126). Its water absorption capacity is very high, affecting the dimension and developing low bond strength with concrete. The structure of bamboo is irregular; thus, the strength may vary drastically. It is a woody grass that can be attacked by termite and decay.

MANHOLE COVER

A manhole cover is a detachable covering on the opening of a manhole. The primary role of this is to prevent accidentally |falling| anyone or anything inside the manhole. Also, it permits the approved individual for repair and maintenance work.

Grade of Manhole Cover

The grade of the manhole cover, its grade designation, and use purpose is shown in Table 1the grade of manhole cover depends on its load resisting capacity and its thickness, as mentioned in IS 12592:2002.

Table 1: Grade of Manhole Cover (IS 12592:2002)

Grade	Grade Designation (Load capacity)	Thickness of manhole as per grade (mm)	Use purpose
Light Duty	LD-2.5 (25KN)	50	Residential and institutional Complexes/areas with pedestrian, inspection chambers
Medium Duty	MD-10 (100KN)	70	Service lanes/roads, on pavements used under medium duty vehicular traffic, car parking areas
Heavy Duty	HD-20 (200KN)	90	Institutional/commercial areas/ carriageways/city trunk roads /bus terminals with heavy duty vehicular traffic of wheel load between 50 to 100KN and parking areas and manhole chambers for road pavement
Extra Heavy Duty	EHD-35 (350KN)	100	Carriageways in commercial/ industrial/port areas/ warehouses/ godowns where frequent loading and unloading of trucks/trailers, slow to fast-moving vehicular traffic for wheel loads up to 115 KN

A significant problem faced in the world is the issue of global warming. The main reason for global warming is due to increasing carbon footprint. The construction sector contributes an extensive amount of carbon footprint from the manufacturing process of steel and cement. Also, due to abundant non-renewable resources used for steel manufacturing, steel prices are rising day by day. Furthermore, as steel reinforced manhole cover and mild steel manhole cover has a good resale value, this manhole cover are in danger of being stolen, which may cause accidents due to open holes left on the road.

OBJECTIVES OF THE INVENTION

The following are the objectives that will be studied in the research work:

1. To explore the use of bamboo as reinforcement in the manhole cover.
2. To carry out an experimental investigation on bamboo reinforced and steel reinforced manhole cover.
3. To compare bamboo reinforced over steel reinforced manhole cover with the following parameter:

- Strength (Load Resisting Capacity)
- Cost
- Carbon footprints

SUMMARY OF THE INVENTION

Materials Used in research work

Cement: The Ordinary Portland cement of 53 grades.

Fine aggregate: Locally available sand passing through 4.75mm sieve.

Coarse aggregate: Locally available crushed blue granite stones aggregate of nominal size 10mm.

Water: Casting and curing of specimens are with portable water that is available on site.

Concrete Admixture: For good workability, water reduction, waterproofing, high performance.

Binding wire: 16-gauge wire for tying reinforcement with each other.

Coat tar epoxy paint: Coal tar epoxy paint for bamboo waterproofing shows in Figure 1.

Figure 1: Coal Tar Epoxy Paint.

Bamboo: The brown color of 3 years old with largest diameter culm.

Fibers: Bamboo fibers shows in Figure 2.

Mould: As per the selected shape and size of the manhole cover.

Figure 2: Bamboo Reinforcement

FLOWCHART OF EXPERIMENTAL WORK

According to the literature, survey bamboo cannot be used directly as reinforcement in concrete. Thus bamboo has to be processed to make it compatible and improve its performance in concrete. Hence the process of selection, seasoning, and treatment is done before use. The flowchart showing the methodology for the preparation process of a bamboo reinforced manhole cover in Figure 3.

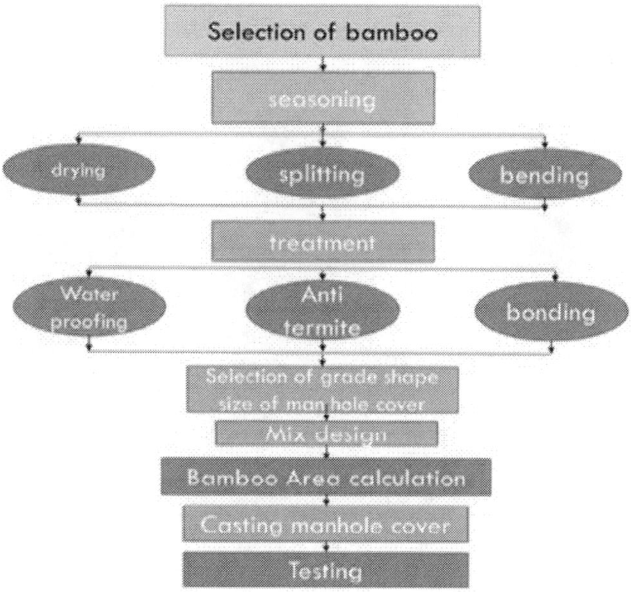

Figure 3: Flowchart of Methodology.

Selection of Bamboo

- Using bamboo as reinforcement in concrete, the following points have to be considered.
- The bamboo appearing brown should be made for use, confirming that the plant is a minimum of three years old.
- Select the longest and largest diameter culms available, which ensure complete growth of culm.
- The complete green, unseasoned bamboo should not be considered for use as these culms are generally weaker due to the high moisture content.

Seasoning of Bamboo

Seasoning is a process that will make bamboo compatible to use as reinforcement in concrete

- **Drying**: the harvested bamboo should be allowed to dry for three to four weeks before use to reduce the moisture content from the fibers
- **Splitting:** the bamboo can be done using a splitting machine or using a sharp knife to obtain straight sectional strips. Figure 4 shows the splitting of bamboo in strips.

Figure 4: Splitting of Bamboo in Strips

- **Bending:** Bamboo strips can be bent with the application of pressure or heat. These bamboo strips can be used as stirrups for additional anchorage in the beam and column.

Treatment of Bamboo

To improve the performance of bamboo as reinforcement, the following treatments are to be carried to get optimal results. Figure 5 shows the treatment done on bamboo strips.

Figure 5: Treatment on Strips of Bamboo with Coal Tar

- **Waterproofing (Low water absorption):** bamboo should be coated with coal tar epoxy paint for low water absorption as bamboo has the capacity to absorb up to 50% of the water of its weight from concrete.
- **Anti-termite:** coating or soaking in lethal solution to resist termite attack.
- **Good bonding in concrete:** a sprinkling of sand on the coated bamboo with coal tar and dried for 24-48 hours. (Ahmad, 2014)

Selection of Grade, Shape and Size of Manhole Cover

The Grade of manhole cover selected in this study light duty, rectangular, 700x550x50mm in dimension and was tested for load resisting capacity of manhole cover.

Mix Design

The grade of concrete selected is M30 as per the recommendation in IS 12592:2002, and for bamboo reinforced, the water-cement ratio must be as low as possible because bamboo can absorb water from concrete, and the tensile strength of bamboo may reduce. Table 2 shows the mix design of M30 grade concrete.

Table 2: Mix Design of M30 Grade

M30 Grade	Materials	kg/m³
1	Cement	380
2	Water	160
3	Fine Aggregate	711
4	Coarse Aggregate	1283
	20 mm coarse Aggregate	924
	10 mm coarse Aggregate	359
5	Admixture	1.9
6	Water Cement Ratio	0.42
7	Bamboo Fibers	1%

Bamboo Area Calculation (Ratio of Bamboo Reinforcement)

For bamboo as reinforcement in concrete, the ratio varies from 2-4% of the area of cross-section. The bamboo reinforcement ratio is taken as 4% of the area of cross-section of manhole cover.

Manual Design Calculation of Manhole Cover

For the design purpose, a manhole cover is considered as a simply supported slab.

Design for Light Duty Bamboo Reinforced Manhole Cover size 700x550x50 mm (4% Bamboo Reinforcement of the Area of Cross-section) figure 6 shows the dimension of manhole cover

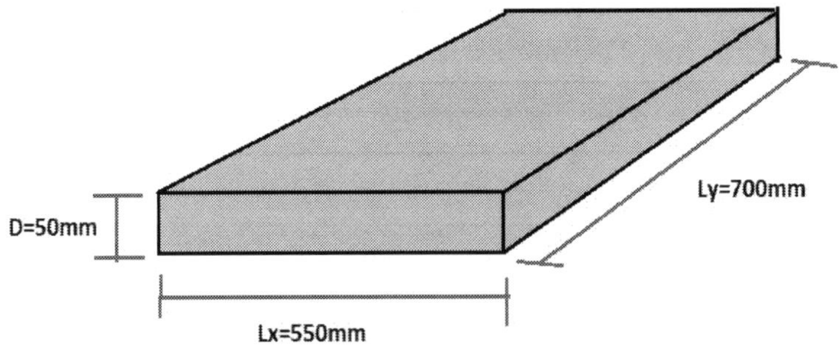

Figure 6: Dimensions of manhole

The grade of concrete taken is M30, and bamboo tensile strength is considered as 125 N/mm2 and bamboo bars assumed to be circular in cross-section as steel bars

Step 1: Type of slab

Ly/Lx = 0.70/0.55 = 1.27< 2 .....(Hence two-way slab)

Ly is the longest span, and Lx is the shortest span of the slab.

Step 2: Depth of slab (IS 456:2000 clause 23.2.1)

d = span / BV * MF = 700 / 20*1.5 = 23.33mm

d is taken as 35mm for further calculation

D = 50mm

BV is the basic value, MF is modification factor, d is effective depth, and D is total depth

Step 3: Effective span (IS 456:2000 clause 22.2)

Leff = length of slab + effective depth (d) = 700 + 35 = 735 mm

Step 4: Load calculation

Dead load = density of RCC* volume = 25*0.0192 = 0.48 KN/m

Live load = target load / length of slab = 25KN / 0.7m = 35.71 KN/m

Total Load = dead load + live load = 36.19 KN/m

Working load (W) =1.5* total load = 1.5 * 36.19 = 54.285 KN/m

Step 5: Ultimate moment (IS 456:2000 clause D-2.1)

αx=0.062, αy=0.062 (IS 456:2000 Table 27)

Mx= $\alpha x * W * Leff^2 = 0.062 * 74.841 * (0.585)^2$ =1.587KNm

My= $\alpha y * W * Leff^2 = 0.062 * 74.841 * (0.735)^2$ = 1.81KNm

Md (ultimate moment) = 1.81 KNm

αx and αy are moment coefficient

Step 6: Check for depth

Xu max = 0.46 d ... (Xu max taken is assumed)

Md = 0.36 * fck * Xu max * b (d- 0.42 Xu max)

$1.81 * 10^6 = 0.36 * 30 * 0.46 d * 1000 * (d - 0.42 * 0.46 d)$

d = 21.25 < 35 mm

fck is compressive strength of concrete

Step 7: Reinforcement calculation

Abr(long span)= (0.5*fck/fy) * (1-(in squareroot(1-((4.6*Md)/fck.

bd²))*bd

= (0.5*30/125)*(1-(insquareroot(1-((4.6*1.81*10⁶)/ 30*1000*35²))*1000*35

Abr (long span)= 506.28 mm ^{2}

Abr(short span) = (0.5*fck/fy) * (1-(in squareroot(1-((4.6*Md)/fck. bd²))*bd

= (0.5*30/125)*(1-(insquareroot(1-((4.6*1.587*10⁶)/ 30*1000*35²))*1000*35

Abr(short span) = 440.22 mm ^{2}

Abr is area of bamboo reinforcement; fy is tensile strength of bamboo (125 N/mm^2).

Step 8: Minimum reinforcement

0.12 % * bd = 0.12 / 100 * 1000 * 35 = 42 mm^2

Step 9: Spacing

Spacing = ((πd²/4) / (Abr required))* 1000 = ((3.14*12²/4) / (506.28))*1000

= 223.38 mm

Hence use 5 number of 12 mm bamboo bars are used in x at 91 mm spacing and 7number of 12 mm bamboo bars are used in x at 87mm spacing center to center.

d is the diameter of bar.

SUMMARY

Grade of concrete used is M30 grade, and approximately 4% of area of cross-section as bamboo reinforcement ratio and cross-sectional area of a single bamboo strip are approximately 113 mm^2. 5 number of 12 mm bamboo bars are used in x at 91 mm spacing and 7number of 12 mm bamboo bars are used in x at 87mm spacing center to center .

Casting of Manhole Cover

For the casting process, the mesh of bamboo strips tied with the binding wire at a required spacing and kept inside the mould (formwork), and then placing concrete with the required amount of tamping is done. Figure 7 shows Bamboo Reinforcement and Figure 8 shows the casting of manhole cover.

Figure 7: Bamboo Reinforcement

Figure 8: Casting of Manhole Covers

Testing on Manhole Cover (Load Resisting Capacity of Manhole cover)

Load test on manhole cover is done in UTM machine, in this test application of load on the manhole cover is done at a rate of 1 KN/s up to its target load with the use of 300mm diameter circular block (plate) placed on its geometrical center as per IS code 12592:2002 recommendation. Figure 9 shows the test setup for a testing manhole cover in UTM.

Figure 9: Test Setup for Testing of Manhole Cover in UTM

PROBLEM STATEMENT

A significant problem faced in the world is the issue of global warming. The main reason for global warming is due to increasing carbon footprint. The construction sector contributes an extensive amount of carbon footprint from the manufacturing process of steel and cement. Also, due to abundant non-renewable resources used for steel manufacturing, steel prices are rising day by day. Furthermore, as steel reinforced manhole cover and mild steel manhole cover has a good resale value, this manhole cover are in danger of being stolen, which may cause accidents due to open holes left on the road.

WE CLAIMS

Our invention gives importance to eco-friendly and natural materials for supporting sustainable development. The non-renewable materials such as steel are being replaced with bamboo which has the potential of getting similar results as steel reinforcement in the concrete according to many studies. The process of selection, seasoning, and treatment of bamboo is carried out before to improve its properties and performance as reinforcement in concrete this research investigates the use of bamboo as reinforcement in the production of reinforced concrete manhole cover as an alternative replacement to steel reinforcement. The main issue with manhole covers is stealing of cast iron and steel reinforced manhole cover which may result in causing accidents due to open holes left on road. The purpose of this study was to investigate the load resisting capacity of bamboo reinforced concrete manhole cover in order to apply this design in construction practice. Testing, grade, and shape are considered as per IS |12592:2002|. This research work makes an approach to develop a new material that would contribute to sustainable development and also green alternative material to steel. This material will no longer become an issue in terms of theft and would contribute towards sustainable development.

- According to claim1# the invention is to the behavior of bamboo reinforcement is similar to steel reinforcement, and the application of bamboo reinforcement in manhole cover will achieve the desirable strength of light duty manhole cover of |+25KN|.
- According to claim1, 2# the invention is to the bamboo reinforcement, the durability of manhole cover will increase as it overcomes the problem of corrosion of steel reinforcement?
- According to claim1, 2, 3# the invention is to the Bamboo reinforcement in manhole cover the cost can be reduced up to |+36%| compared to steel reinforced manhole cover.
- According to claim1, 2, 4# the invention is to the carbon footprint emission can be reduced up to |+43%| using bamboo reinforcement in manhole cover, which can lower the effect of global warming and contribute towards sustainable development.

CITATIONS

- Miss. Madhulika Sinha, Adith Selvaraj, Vishnu Panicker, Amol Niladhe, Prathamesh Pawar, 202121062302, MAHATMA EDUCATION SOCIETY'S, PILLAI HOC COLLEGE OF ENGINEERING &TECHNOLOGY RASAYANI, Paver blocks of Concrete with Recycled Aggregates and Partial Replacement of Cement with Alccofine", 2022, Indian Patent office.
- NAIR ROHAN UNNIKRISHNAN, 202221000050, "प्राबल्यम-ASCENDING" MAHATMA EDUCATION SOCIETY'S, PILLAI COLLEGE OF ENGINEERING, 2022, Indian Patent office.
- Ms. CHAITALI R. KULKARNI, Mr. RAJU NARWADE, Mr. KARTHIK NAGARAJAN, DR. G R PATIL, 202221000049, " Smart Eco-friendly E-waste Divider Block (SEED)" MAHATMA EDUCATION SOCIETY'S, PILLAI HOC COLLEGE OF ENGINEERING &TECHNOLOGY RASAYANI, 2022, Indian Patent office.
- Dr. Daphne Pillai, 2022210010567,Director, Pillai College of Arts, Commerce and Science," New methods of education to meet post-Covid requirements" MES's Pillai College of Arts, Commerce and Science, Dr. K. M. Vasudevan Pillai Campus, Sector 16, New Panvel, 401206 Navi Mumbai, India, 2022, Indian Patent office
- 202021106920, Granted, Dr. Sharda Sharma, Mathoscope, MAHATMA EDUCATION SOCIETY'S, Dr. Pillai Global Academy, Plot no.1, RSC-48, Gorai 2, Borivali (West), Mumbai, 400092.
- 202022100309, Granted, Valectus Private Limited, Method for Storage and Processing of Geospatial Data for Efficient Rendering and Visualization on a Server, : 18-B, Gurudev Apartments, 1st Floor, Chembur Naka, R C Marg, Chembur, Mumbai, Mumbai City, Maharashtra, India, 400074.
- 202021106919, Granted, Dr. Daphne Pillai, Dr. Priam Pillai, Franav Pillai, Deepika Sharma, Dr. Kavita Kathare, Skill Credit System, MAHATMA EDUCATION SOCIETY'S, Pillai College of Arts Commerce & Science(Autonomous)
- 202221001647, MAHATMA EDUCATION SOCIETY'S, Pillai College of Arts Commerce & Science, White Board "Dry and Wet" Flip Duster Eraser" Dr. Gajanan Wader, Principal, Dr. Shabab Ahmed Rizvi, Assistant Professor.
- 202221001646, Dr. Aarti Sukheja, Dr. Kavita Kathare, Intelligent Air Conditioning System Notify The part that needs repair or replacement, MAHATMA EDUCATION SOCIETY'S, Pillai College of Arts, Commerce & Science (Autonomous).

- 202221001645, Dr. G R Patil, Dr. K D Singh,Prof. Raju Narwade, Prof. Karthik Nagarajan., Prof. Manisha Jamgade Dual triangular slope bottom tuned liquid damper for improving efficiency of flat bottom tuned liquid damper. MAHATMA EDUCATION SOCIETY'S, PILLAI HOC COLLEGE OF ENGINEERING & TECHNOLOGY RASAYANI.

CHAPTER - 08

Sustainable Sandwich Composite Circular Manhole Cover Designed for Pedestrian Networks

Sonam Prasad Bhopatrao, Prof (Dr.) Shilpa Pankaj Kewate

Mahatma Education Society's, Pillai Hoc College of Engineering & Technology Rasayani Taluka Panvel, Dist, Navi Mumbai, Maharashtra 410207, India.

931 Prasad bungalow, At & Post- Kadav, Tal- Karjat, District- Raigad, Pin-410201, MH, India.

ABSTRACT

The motive of this invention is basically promoting the use of natural as well as sustainable materials. In most developing countries, the demand for conventional construction materials leads to the overexploitation of natural resources required for their manufacturing along with their adverse effects on the environment. Such as steel, which is the most consumed material because of its high tensile strength. But this is very costly and also affects the environment badly as it produces a sufficient amount of carbon gases to pollute the environment during its manufacturing process. And to cope with this situation, bamboo is used as a substitute for steel reinforcement in the manhole cover. Also, the possibility of being stolen is increases due to its good amount of scrap value and if holes will remain open due to stealing of covers there will be a chance of accident. This research work is focused on replacing the steel reinforcement with bamboo which is naturally available, highly renewable and strong in both tension as well as compression. Various types of treatment methods used which results the good potential of bamboo for the replacement of steel reinforcement by achieving remarkable bond strength corresponding to steel reinforcement. Fine aggregate is partially replaced by 7.5% with the E-waste powder which will promote the strength of the product. This addition of E-waste will give multiple benefits as it conserves the natural resources, resolving the problems facing in the disposal of E-waste and ultimately saving the lands that are used for disposal purposes and can prevent the ecosystem to be polluted. Although there are shapes available in manholes like squares or rectangles then also the Circular shape seems to be a good shape of a manhole as it sustains the compression of earth. A circular shape of a manhole cover can be proved ideal to manufacture, that's why this invention is basically about to work on a Circular shaped manhole cover. In this work, the bamboo and chicken mesh are used to make the sandwich composite Circular manhole cover by using the middle layer of bamboo reinforcement having two chicken meshes along both the sides of bamboo embedded in concrete. There will be a good impact of this product on cost and environmental aspects. This type of Circular manhole covers surely contribute towards eco-friendly and sustainable development as well as will help to preserve the natural resources.

FIELD OF THE INVENTION

This invention is related to Sustainable Circular Manhole cover with Bamboo as Reinforcement for pedestrian networks.

BACKGROUND OF THE INVENTION

Although there are shapes available in manholes like squares or rectangles then also the Circular shape seems to be a good shape of a manhole as it sustains the compression of earth. The procedure of the manufacturing of circular shape manhole is quite easy than the square or rectangular shape. Sufficient efforts are needed in the installation of manholes as they are quite heavy but if we are selecting the shape of the manhole as circular instead of square or rectangle, it will reduce the efforts as one can roll it to the required places. In the case of a circular manhole cover, there will be fewer chances of falling through an opening as compared to square or rectangular manholes because there is a possibility of the square manhole falling in due to the larger diagonal length of the opening. As compared to the width of the square manhole, a round-shaped manhole can save the material used for casting without changing its width as that of square shape hence ultimately resulting in low cost. Thus, a circular shape of a manhole cover can be proved ideal to manufacture.

As we can see concrete has good compressive strength but it is low in tensile strength. To cope up with this condition in concrete structures, steel is introduced as reinforcement in concrete for promoting the tensile strength required for the durability of the structure. However, some issues commonly faced with steel are increment in overall cost, use of the non-renewable resources for its manufacturing, polluting environment due to industrial processing, a significant contributor for the overall dead load of structures, etc. Therefore, sustainable construction processes are thus required which involves the application of materials that will take care of the environment & sustainability too.

Steel reinforced manhole cover, and mild steel manhole cover has a good scrap value due to which there is the possibility of manhole cover to being stolen is increases and if holes will remain open due to stealing of covers there will be a chance of accidents which can sometimes lead to death. Accidents can happen in case of public as well as animals can also fall down in the manhole while walking on the road. Again, the open holes of the manhole may disturb the lives of residing people due to foul odors and arising to various diseases in case of sewer manholes.

For manufacturing 1 ton of steel, there is an emission of approximately 2 tons of carbon dioxide, and for manufacturing 1 ton of cement, there is an emission of approximately 1 ton of carbon dioxide in the environment. Due to the increase of carbon footprint in the environment, the earth's temperature rises year by year. If this situation continues, it will cause an unfavorable environment for humans and other living organisms, and the sustainability of life will be in danger.

As there is broad industrialization globally, the ecosystem is affected badly due to the release of dangerous gases in the environment from various manufacturing sectors. And this is ultimately results in unfavorable conditions such as unpredictable rainfall, rise in sea level, global warming, acidification of oceans, earthquakes, etc. and also the crucial problem encountered by the construction sector is carbon emission from steel and cement manufacturing. The construction sector contributes up to 11% of carbon dioxide emissions globally.

BAMBOO

Bamboo is a part of the grass family of Poaceae, which belongs to the sub-family of Bambusoideae. It is a combined element made up of lengthy and parallel cellulose fibers compressed in a ligneous matrix form. There is an abundant number of species of bamboo found in the whole world. The carbon sink by 5-year-old bamboo every year is 18 to 23 tons per acre.

Upsides of Bamboo

- Naturally available everywhere.
- Light weight compared to steel.

- Low-cost
- Bio degradable
- Bamboo can prevent pollution by absorbing large amounts of nitrogen from waste water.
- Reducing the amount of carbon dioxide in the air.

Downsides of Bamboo

- Tensile strength less than steel
- High water absorption due to which it swells
- Prone to termite attack and rotting
- Non uniform structure
- Low bond strength than steel with concrete

MANHOLE COVER

A manhole is an adequate hole in a path or road, covered by a removable lid that can be helpful to the workers or inspecting person to get the access for maintenance, inspection and system upgrades. This product is also the best solution to avoid the accidental cases happened due to the opening of manhole remains uncovered.

Grade of Manhole Cover

The grade of the manhole cover, its grade designation, and use purpose is shown in Table 1the grade of manhole cover depends on its load resisting capacity and its thickness, as mentioned in IS 12592:2002.

Table 1: Grade of Manhole Cover (IS 12592:2002)

Grade	Grade Designation (Load capacity)	Thickness of manhole as per grade (mm)	Use purpose
Light Duty	LD-2.5 (25KN)	50	Residential and institutional Complexes/areas with pedestrian, inspection chambers
Medium Duty	MD-10 (100KN)	70	Service lanes/roads, on pavements used under medium duty vehicular traffic, car parking areas

Grade	Grade Designation (Load capacity)	Thickness of manhole as per grade (mm)	Use purpose
Heavy Duty	HD-20 (200KN)	90	Institutional/commercial areas/ carriageways/city trunk roads / bus terminals with heavy duty vehicular traffic of wheel load between 50 to 100KN and parking areas and manhole chambers for road pavement
Extra Heavy Duty	EHD-35 (350KN)	100	Carriageways in commercial/ industrial/port areas/ warehouses/ godowns where frequent loading and unloading of trucks/trailers, slow to fast-moving vehicular traffic for wheel loads up to 115 KN

OBJECTIVES OF THE INVENTION

The following are the objectives that will be studied in the research work:

1. To use the better shape of a manhole cover, that is a 'Circular' which will be beneficial in various ways.
2. To explore the use of bamboo as reinforcement in the manhole cover.
3. To carry out an experimental investigation on bamboo reinforced manhole cover.
4. To analyze the properties of bamboo reinforced manhole cover with the following parameter:
    - Strength (Load Resisting Capacity)
    - Cost
    - Carbon footprints

SUMMARY OF THE INVENTION

Materials Used in research work

- **Cement:** The Ordinary Portland cement of 53 grades.
- **Fine aggregate:** Locally available sand passing through 4.75mm sieve.
- **E-Waste Powder** as a partial replacement.

- **Coarse aggregate:** Locally available crushed blue granite stones aggregate of nominal size 10mm.
- **Water:** Casting and curing of specimens are with portable water that is available on site.
- **Concrete Admixture:** For good workability, water reduction, waterproofing, high performance.
- **Binding wire:** 16-gauge wire for tying reinforcement with each other.

Sikadur 32 gel. Figure 3.

Figure 3: Sikadur 32 Gel.

Bamboo: The brown color of 3 years old with largest diameter culm.

Fibers: Bamboo fibers shows in Figure 2.

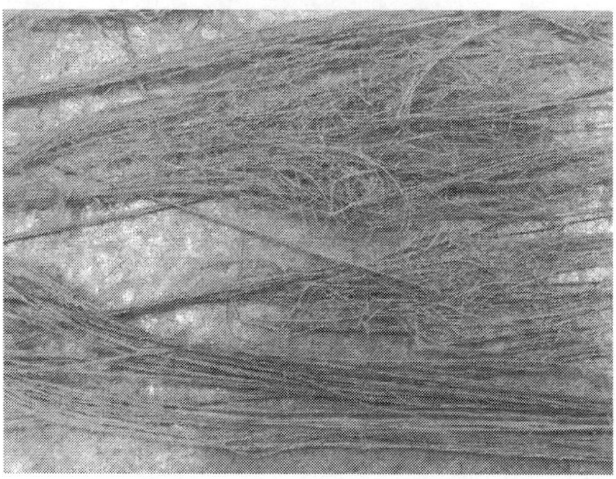

Figure 5: Bamboo Fibres.

E-Waste Powder as a replacement of fine aggregate. Figure 7.

Figure 7: E-Waste Powder

Chicken Mesh. Figure 8.

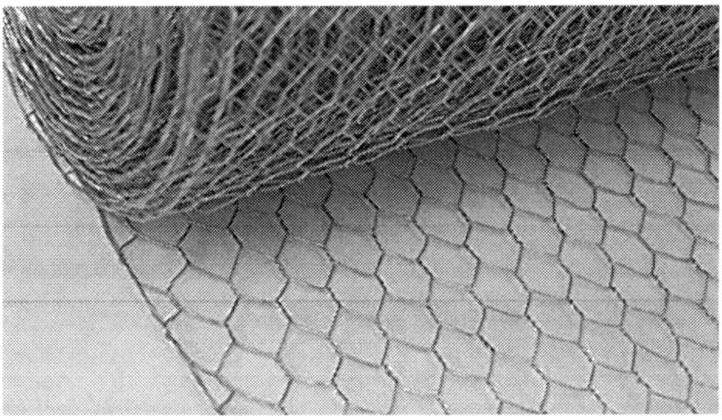

Figure 8: Chicken Mesh

Mould: As per the selected shape and size of the manhole cover.

DESCRIPTION OF THE DIAGRAM

Fig. 1: Flowchart of Methodology.

Fig. 2: Splitting of Bamboo in Strips.

Fig. 3: Sikadur 32 Gel.

Fig. 4: Treatment on Strips of Bamboo with Sikadur 32 Gel.

Fig. 5: Bamboo Fibers.

Fig. 6: Bamboo Reinforcement.

BRIEF DESCRIPTION OF THE DIAGRAM

Flowchart of Experimental Work

According to the literature, survey bamboo cannot be used directly as reinforcement in concrete. Thus, bamboo has to be processed to make it compatible and improve its performance in concrete. Hence the process of selection, seasoning, and treatment is done before use. The flowchart showing the methodology for the preparation process of a bamboo reinforced manhole cover in Figure 1.

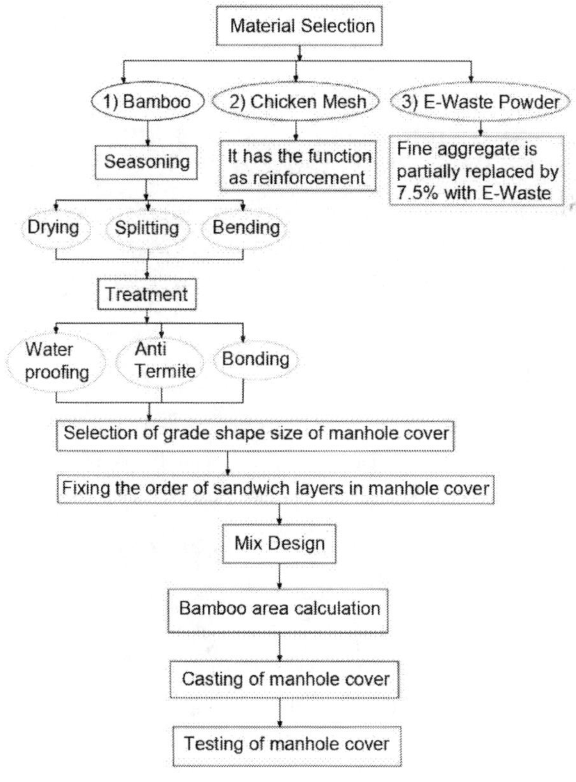

Figure 1: Flowchart of Methodology.

Selection of Bamboo

- Using bamboo as reinforcement in concrete, the following points have to be considered.
- The bamboo appearing brown should be made for use, confirming that the plant is a minimum of three years old.
- Select the longest and largest diameter culms available, which ensure complete growth of culm.
- The complete green, unseasoned bamboo should not be considered for use as these culms are generally weaker due to the high moisture content.

Seasoning of Bamboo

Seasoning is a process that will make bamboo compatible to use as reinforcement in concrete

- **Drying**: the harvested bamboo should be allowed to dry for three to four weeks before use to reduce the moisture content from the fibers
- **Splitting:** the bamboo can be done using a splitting machine or using a sharp knife to obtain straight sectional strips. Figure 2 shows the splitting of bamboo in strips.
- **Bending:** Bamboo strips can be bent with the application of pressure or heat. These bamboo strips can be used as an outer ring as a circumferential reinforcement to avoid slipping at the edges.

Treatment of Bamboo

To improve the performance of bamboo as reinforcement, the following treatments are to be carried to get optimal results. Figure 4 shows the treatment done on bamboo strips.

Figure 4: Treatment on Strips of Bamboo with Sikadur 32 Gel.

- **Waterproofing (Low water absorption):** bamboo should be coated with sikadur 32 gel for low water absorption as bamboo has the capacity to absorb up to 50% of the water of its weight from concrete.
- **Anti-termite:** coating or soaking in lethal solution to resist termite attack.
- **Good bonding in concrete:** a sprinkling of sand on the coated bamboo with sikadur 32 gel and dried for 24-48 hours.

Selection of Grade, Shape and Size of Manhole Cover

The Grade of manhole cover selected in this study light duty, circular, 560x50mm in dimension and was tested for load resisting capacity of manhole cover.

Mix Design

The grade of concrete selected is M30 as per the recommendation in IS 12592:2002, and for bamboo reinforced, the water-cement ratio must be as low as possible because bamboo can absorb water from concrete, and the tensile strength of bamboo may reduce. Table 2 shows the mix design of M30 grade concrete.

Table 2: Mix Design of M30 Grade

M30 Grade	Materials	kg/m^3
1	Cement	360
2	Water	144
3	Fine Aggregate	812
	Sand	751.10
	E-Waste Powder (7.5%)	60.90
4	Coarse Aggregate	1284
	20 mm coarse Aggregate	645
	10 mm coarse Aggregate	639
5	Admixture	1.8
6	Water Cement Ratio	0.40
7	Bamboo Fibers	1%

Bamboo Area Calculation (Ratio of Bamboo Reinforcement)

For bamboo as reinforcement in concrete, the ratio varies from 2-4% of the area of cross-section. The bamboo reinforcement ratio is taken as 4% of the area of cross-section of manhole cover.

Manual Design Calculation of Manhole Cover

For the design purpose, a manhole cover is considered as a simply supported slab.

Design for Light Duty Bamboo Reinforced Manhole Cover size 560x50 mm (4% Bamboo Reinforcement of the Area of Cross-section) figure 9 shows the dimension of manhole cover

The grade of concrete taken is M30, and bamboo tensile strength is considered as 125 N/mm2 and bamboo bars assumed to be circular in cross-section as steel bars

Step 1: Type of slab

Bending of a circular slab like a saucer (i.e., bending in all directions) is quite different form bending in one or two distinct orthogonal directions for a rectangular slab under uniformly distributed loads. In other words, the stresses caused in circular slabs are radial and circumferential necessitating reinforcements in these directions. Provision of such reinforcements becomes difficult due to congestion in the central portion of the slab where maximum bending moments in both directions occur. Hence, in practice, reinforcements are provided in two orthogonal directions as those for two-way rectangular slabs.

Step 2: Depth of slab (IS 456:2000 clause 23.2.1)

d = span / BV * MF = 560 / 20*1.5 = 18.66mm

d is taken as 35mm for further calculation

D = 50mm

BV is the basic value, MF is modification factor, d is effective depth, and D is total depth

Step 3: Effective span (IS 456:2000 clause 22.2)

Leff = length of slab + effective depth (d) = 560 + 35 = 595 mm

Step 4: Load calculation

Dead load = density of RCC* volume = 25*0.0123 = 0.31 KN/m

Live load = target load / length of slab = 25KN / 0.56m = 44.64 KN/m

Total Load = dead load + live load = 44.95 KN/m

Working load (W) =1.5* total load = 1.5 * 44.95 = 67.425 KN/m

Step 5: Ultimate moment (IS 456:2000 clause D-2.1)

Radial and circumferential B.M. at centre per meter width of slab:

$(M_r)_c = (M_\theta)_c = 3/16 * Wa^2 = 3/16 * (67.425) * (0.280)^2 = 0.99$ KNm

Circumferential B.M. at support per meter width of slab

(Mθ) e = 2/16 *Wa^2 = 2/16* (67.425) * (0.280)^2 = 0.66 KNm

Radial B.M. at support per meter width of slab, (Mr) e = 0

Design Bending Moment = 0.99 KNm

Step 6: Check for depth

Xu max = 0.46 d ... (Xu max taken is assumed)

Md = 0.36 * fck * Xu max * b (d- 0.42 Xu max)

0.99 * 10^6 = 0.36 * 30 * 0.46 d *1000 *(d- 0.42*0.46 d)

d = 15.78 < 35 mm

fck is compressive strength of concrete

Step 7: Reinforcement calculation

Abr = (0.5*fck/fy) * (1-(in squareroot(1-((4.6*Md)/fck.bd^2))*bd

= (0.5*30/125) *(1-(insquareroot(1-((4.6*0.99*10^6)/

30*1000*35^2))*1000*35

Abr = 268.83 mm ^{2}

Abr is area of bamboo reinforcement; fy is tensile strength of bamboo (125 N/mm^2).

Step 8: Minimum reinforcement

0.12 % * bd = 0.12 / 100 * 1000 * 35 = 42 mm^2

Step 9: Spacing

Spacing = ((пd^2/4) / (Abr required))* 1000 = ((3.14*12^2/4) / (268.83))*1000

= 420.33 mm

Hence use 5 number of 12 mm bamboo bars are used in both the direction at 90 mm spacing center to center.

d is the diameter of bar.

To avoid slipping at the edges, provide extra circumferential

reinforcement in the form of rings.

Provided 2 rings of 12mm bamboo @ 40 mm c/c at the edge

SUMMARY

Grade of concrete used is M30 grade, and approximately 4% of area of cross-section as bamboo reinforcement ratio and cross-sectional area of a single bamboo strip are approximately 113 mm^2. 5 number of

12 mm bamboo bars are used in both the direction at 90 mm spacing center to center and 2 rings are provided at edges @ 40mm c/c. Figure 6 shows bamboo reinforcement.

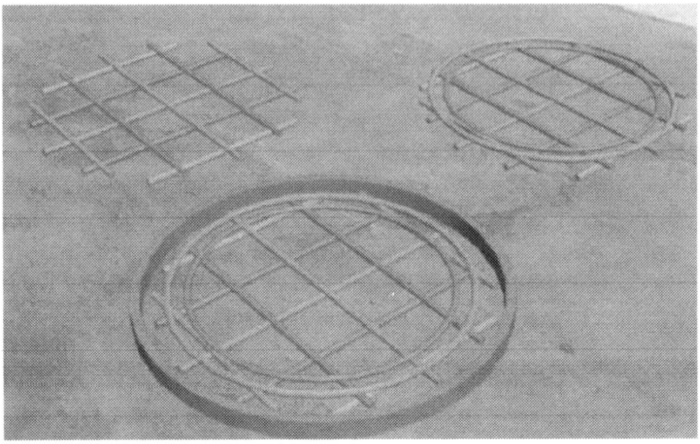

Figure 6: Bamboo Reinforcement

Casting of Manhole Cover

For the casting process, the mesh of bamboo strips tied with the binding wire at a required spacing and kept inside the mould (formwork), and to the both side of bamboo, chicken mesh is provided and then placing concrete with the required amount of tamping is done. Figure 10. show the casting of manhole cover.

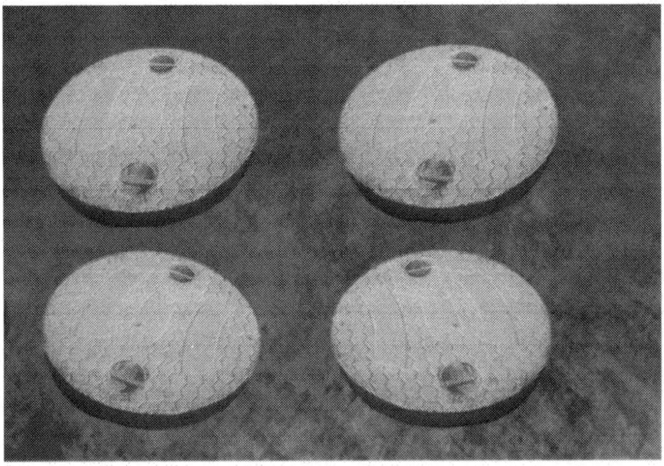

Figure 10: Casting of Manhole Covers

Testing on Manhole Cover (Load Resisting Capacity of Manhole Cover)

Load test on manhole cover is done in UTM machine, in this test application of load on the manhole cover is done at a rate of 1 KN/s up to its target load with the use of 300mm diameter circular block (plate) placed on its geometrical centre as per IS code 12592:2002 recommendation. Figure 11. shows the test setup for a testing manhole cover in UTM.

Figure 11: Test Setup for Testing of Manhole Cover in UTM

PROBLEM STATEMENT

Non one can assure that an up-coming generation will experience an eco-friendly environment, as we all are facing the major issue of global warming overall the world. Increased carbon emission is one of the main reasons of global warming. Steel and cement are widely used materials in construction industry and are releasing more and more carbon gases in the environment during their manufacturing process. Along with the pollution one more downside of the steel is it's day by day rising cost. In addition to this, as steel reinforced manhole cover and mild steel manhole cover has good amount of scrap value, there will be a possibility of manhole cover to being stolen which may leads to the accidents if manhole will remain open on the road.

WE CLAIMS

Our invention gives importance to natural and eco-friendly materials for supporting sustainable development. The non-renewable materials such as steel are being replaced with bamboo which has the potential of getting similar results as steel reinforcement in the concrete according to many studies. The process of selection, seasoning, and treatment of bamboo is carried out before to improve its properties and performance as reinforcement in concrete this research investigates the use of bamboo as reinforcement in the production of reinforced concrete manhole cover as an alternative replacement to steel reinforcement. The main issue with manhole covers is stealing of cast iron and steel reinforced manhole cover which may result in causing accidents due to open holes left on road. The purpose of this study was to investigate the load resisting capacity of bamboo reinforced concrete manhole cover in order to apply this design in construction practice. Testing, grade, and shape are considered as per IS |12592:2002|. This research work makes an approach to develop a new material that would contribute to sustainable development and also green alternative material to steel. This material will no longer become an issue in terms of theft and would contribute towards sustainable development.

- According to claim1# the invention is to the behaviour of bamboo reinforcement is similar to steel reinforcement, and the application of bamboo reinforcement in manhole cover will achieve the desirable strength of light duty manhole cover of |+25KN|.
- According to claim1, 2# the invention is to the bamboo reinforcement, the durability of manhole cover will increase as it overcomes the problem of corrosion of steel reinforcement.
- According to claim1, 2, 3# the invention is to the Bamboo reinforcement in manhole cover the cost can be reduced as compared to steel reinforced manhole cover.
- According to claim1, 2, 4# the invention is to the carbon footprint emission can be reduced using bamboo reinforcement in manhole cover, which can lower the effect of global warming and contribute towards sustainable development.

CITATIONS

- 202221002498, SWEEKRITI ASSET LLP, Dr. Tusharika Sinha Banerjee, Dr. Arti Hadap,Advanced Vacuum Electron Tubes to get Higher Output Power., Faculty, Department of Electronics and Telecommunication Engineering,Pillai College of Engineering,Dr. K. M. Vasudevan Pillais' Campus Plot No. 10, Sector 16, New Panvel, Dist. Raigad
- 202221006894, Meera Ranjith, Sidra Dalvi, Steffi Misquitta, Manisha Vinod Prasad, Pinky Ravi Shivdasani Augmented Reality in Teacher Education Curriculum in Mumbai for Enhancing Interest and Retention of Geography Concepts., MES, Mahatma Junior College of Teacher Education, Opposite Fire Brigade, Sion Trombay Road, Chembur Naka, Mumbai 400071, India.
- 202221006891, MES's Mahatma Night Degree College of Arts and Commerce, Customer Relationship Management and its Reflection on Level of Customers' Satisfaction. Dr. (Mrs.) Anaya Aditya Markandeya Dr.(Mrs.) Kishori Jagdish Bhagat
- 202221006890, MES's Mahatma Night Degree College of Arts and Commerce, Strategies of Banks and Significance of Non-Interest Income of Banking Sector in India., Dr. Eknath Kundlik Zhrekar, Prof. Dr. Elizabeth Mathews
- 202221006893, Esakkiammal Sudalai Pandi Nadar, MUMBAI BEST BUS TRACKER, MES, Mahatma Junior College of Teacher Education, Opposite, Fire Brigade, Sion Trombay Road, Chembur Naka, Mumbai 400071, India.
- 202221008799, Dr. Ashish Tripathi, Ms. Shilpa Tripathi, Ms. Agrima Tripathi, Without any Software I Have Built an Automated ATTENDANCE System, MAHATMA EDUCATION SOCIETY'S, PILLAI COLLEGE OF ENGINEERING Management,
- 202221010401, Dr. Mansi Subhedar,Dr. B. K. Sarkar, LED Dimming Driver Circuit Using PWM MAHATMA EDUCATION SOCIETY'S, PILLAI HOC COLLEGE OF ENGINEERING &TECHNOLOGY RASAYANI, TALUKA PANVEL,
- 202221010400, Prof. (Dr.) Shilpa Pankaj Kewate, 3 Bit Error-Correcting Modified Linear Block Code MAHATMA EDUCATION SOCIETY'S, PILLAI HOC COLLEGE OF ENGINEERING &TECHNOLOGY RASAYANI, TALUKA PANVEL,
- 202241010373, Dr. B. K. Sarkar, TBAR-Managements System: Tourist Behaviour Analysis and Advanced Business Managements System, MAHATMA EDUCATION SOCIETY'S, PILLAI HOC COLLEGE OF ENGINEERING &TECHNOLOGY RASAYANI, TALUKA PANVEL

- 202211008805, Prof.(Dr.) Biplab Kumar Sarkar, Normal Mode Analysis and Management Molecular Docking and Structure-Based Drug Design Strategies, MAHATMA EDUCATION SOCIETY'S, PILLAI HOC COLLEGE OF ENGINEERING &TECHNOLOGY RASAYANI, TALUKA PANVEL
- 202221008802, Prof.(Dr.) Biplab Kumar Sarkar, Design Develop and Management of Solar Concentrator studded with optical fiber for Daylighting, MAHATMA EDUCATION SOCIETY'S, PILLAI HOC COLLEGE OF ENGINEERING &TECHNOLOGY RASAYANI, TALUKA PANVEL
- 202231007537, Dr. Biplab Kumar Sarkar, Helmet and License Plate Detection Using Machine Learning and AI- Based Programming., MAHATMA EDUCATION SOCIETY'S, PILLAI HOC COLLEGE OF ENGINEERING &TECHNOLOGY RASAYANI TALUKA PANVEL
- 202221004892, Dr. Biplab Kumar Sarkar, Point Regression Analysis-Based Big Data Processing Using Machine Learning and Deep Learning., MAHATMA EDUCATION SOCIETY'S, PILLAI HOC COLLEGE OF ENGINEERING &TECHNOLOGY RASAYANI, TALUKA PANVEL

Chapter - 09

Experimental Studies on the Use of Waste Rubber Tyre Crumb and Bamboo in Tremix Concrete Flooring

Amar Yashwant Malabari, Prof (Dr.) Shilpa Pankaj Kewate

Mahatma Education Society's, Pillai Hoc College Of Engineering &Technology Rasayani Taluka Panvel, Dist, Navi Mumbai, Maharashtra 410207, India.

ABSTRACT

Water is the fundamental constituent of the solid. As the water concrete proportion influences the quality of the solid however for the ground surface, the main PCC is to be laid on a plinth and afterward, tiles are conferred on PCC work for a better appearance. In any case, if there should be an occurrence of industrial flooring, just PCC work to be done however not tiles to be laid on the floor. For the better appearance of the floor of the business assembling the vacuum dewatering process is utilized from which the water is expelled from the solid for the settlement. This strategy is utilized in private structures for the apparent reason just as forbear the heap like strolling of individual, home apparatuses and so on this kind of deck is utilized in the section of quake safe structure which continues the ground shaking vibration happens because of the seismic tremor. The thickness of the deck is 200 to 300mm which is utilized in a chunk of 5 story construction yet due to vibration that kind of section doesn't to be fell. This sort of ground surface is affordable in light of the fact that the tiles cost to be diminished.

The motive of this invention is basically to promote the use of natural as well as sustainable materials. In most developing countries, the demand for conventional construction materials leads to the overexploitation of natural resources required for their manufacturing along with their adverse effects on the environment.

Such as steel, which is the most consumed material because of its high tensile strength. But this is very costly and also affects the environment badly as it produces a sufficient amount of carbon gases to pollute the environment during its manufacturing process. And to cope with this situation, bamboo is used as a substitute for steel reinforcement in the manhole cover. This research work is focused on replacing the steel reinforcement with bamboo which is naturally available, highly renewable, and strong in both tensions as well as compression. Various types of treatment methods are used which result from the good potential of bamboo for the replacement of steel reinforcement by achieving remarkable bond strength corresponding to steel reinforcement. Fine aggregate is partially replaced by 10% with the Waste rubber tyre crumb which will promote the strength of the product. This addition of a Waste rubber tyre crumb will give multiple benefits as it conserves the natural resources, resolving the problems facing in the disposal of Waste rubber tyre crumb and ultimately saving the lands that are used for disposal purposes and can prevent the ecosystem to be polluted. This type of Tremix Concrete Flooring surely contributes towards eco-friendly and sustainable development as well as will help to preserve the natural resources.

FIELD OF THE INVENTION

This invention is related to Sustainable Tremix concrete flooring with Bamboo as Reinforcement and Waste Tyre Rubber Crumb as 10% replacement for Fine Aggregate.

BACKGROUND OF THE INVENTION

"A Tremix strategy can be characterized as "Vacuum Dewatered Flooring is an exceptional kind of Flooring Technique to accomplish High Strength, Longer Life, Better Finish and Faster Work. This kind of floor is appropriate for high scraped spots and substantial traffic development."

A great solid floor or asphalt requires not exclusively to be level, yet it ought to likewise have high wear obstruction, high compressive quality, decreased shrinkage, and least water porousness.

The Tremix Concrete Flooring technique is a framework for laying great solid floors with predominant cost-adequacy. The way to the utilization of this technique is the dewatering of cement by vacuum process. Surplus water from the solid is expelled quickly, subsequent to setting and vibration, decreasing the water-concrete proportion to an ideal level. As we can see concrete has good compressive strength but it is low in tensile strength. To cope up with this condition in concrete structures, steel is introduced as reinforcement in concrete for promoting the tensile strength required for the durability of the structure. However, some issues commonly faced with steel are increment in overall cost, use of the non-renewable resources for its manufacturing, polluting environment due to industrial processing, a significant contributor for the overall dead load of structures, etc. Therefore, sustainable construction processes are thus required which involves the application of materials that will take care of the environment & sustainability too.

The vacuum dewatering process evacuates surplus water present in the solid. This is finished utilizing the Vacuum Equipment involving Suction Mat Top Cover, Filter cushions, and Vacuum Pump. The procedure begins following surface vibration.

Filter cushions are set on the crisp solid leaving around 4 creeps of new cement uncovered on all sides. The Top Cover is then set on the channel cushions and turned out till it covers the pieces of uncovered cement on all sides. The Top Cover is then associated with the vacuum siphon through a suction hose and the siphon is begun. Vacuum is promptly made between the channel cushions and the top spread. Environmental weight packs the solid and the surplus water is crushed out. This procedure brings down the water content in the solid by 15-25%. The dewatering activity takes roughly 1.5 - 2 minutes for every centimeter thickness of the floor. The dewatered concrete is compacted and dried to such a degree, that it is conceivable to stroll on it without leaving any impressions. This is the sign of cement being appropriately dewatered and prepared for wrapping up.

For manufacturing 1 ton of steel, there is an emission of approximately 2 tons of carbon dioxide, and for manufacturing 1 ton of cement, there is an emission of approximately 1 ton of carbon dioxide in the environment. Due to the increase of carbon footprint in the environment, the earth's temperature rises year by year. If this situation continues, it will cause an unfavorable environment for humans and other living organisms, and the sustainability of life will be in danger.

As there is broad industrialization globally, the ecosystem is affected badly due to the release of dangerous gases in the environment from various manufacturing sectors. And this ultimately results in unfavorable conditions such as unpredictable rainfall, rise in sea level, global warming, acidification of oceans, earthquakes, etc., and also the crucial problem encountered by the construction sector is carbon emission from steel and cement manufacturing. The construction sector contributes up to 11% of carbon dioxide emissions globally.

BAMBOO

Bamboo is a part of the grass family of Poaceae, which belongs to the sub-family of Bambusoideae. It is a combined element made up of lengthy and parallel cellulose fibers compressed in a ligneous matrix form. There is an abundant number of species of bamboo found in the whole world. The carbon sink by 5-year-old bamboo every year is 18 to 23 tons per acre.

Upsides of Bamboo

- Naturally available everywhere.
- Lightweight compared to steel.
- Low-cost
- Biodegradable
- Bamboo can prevent pollution by absorbing large amounts of nitrogen from wastewater.
- Reducing the amount of carbon dioxide in the air.

Downsides of Bamboo

- Tensile strength less than steel
- High water absorption due to which it swells
- Prone to termite attack and rotting
- Non-uniform structure
- Low bond strength than steel with concrete

TREMIX CONCRETE FLOORING

Tremix Concrete Flooring is also known as Vacuum Dewatered Flooring which is a special type of Flooring technique to achieve high strength, longer life, better finish and faster work. This type of flooring is suitable for high abrasion and heavy traffic movement. The Tremix concrete flooring is a system for laying high-quality concrete floors where the key is Dewatering of concrete by using a vacuum pump.

Grade of Tremix Concrete Flooring

The grade of the manhole cover, its grade designation, and use purpose is shown in Table 1 the grade of manhole cover depends on its load resisting capacity and its thickness, as mentioned in IS 12592:2002.

Table 1: Grade of Tremix Concrete Flooring (IS 12592:2002)

Grade	Grade Designation (Load capacity)	Thickness of Tremix concrete floor as per grade (mm)	Use purpose
Light Duty	LD-2.5 (25KN)	200	Residential and institutional Complexes/areas with Pathways, concrete flooring
Heavy Duty	HD-20 (200KN)	300	Industrial Floors/Concrete roads /Runways/Railways Platforms/ Warehouse container Yards/Canal Lining/Bridges/RCC Roads with heavy duty vehicular traffic of wheel load between 50 to 100KN and parking areas

OBJECTIVES OF THE INVENTION

The following are the objectives that will be studied in the research work:

1. To productively expel abundance water from recently set, compacted, and leveled solid surfaces.
2. To improve concrete flooring physical behavior.
3. To get good quality & to achieve speed in construction of concrete floors.
4. To increase in compressive quality of cement by 40-70%.
5. To minimize crack formation, shrinkage was reduced by 50%.
6. To improve wear obstruction.
7. To high and early quality, limit harm to recently cast floors.
8. To analyze the properties of bamboo reinforced tremix concrete floor with the following parameter:
    - Strength (Load Resisting Capacity)
    - Cost
    - Carbon footprints

SUMMARY OF THE INVENTION

Materials Used in research work:

- **Cement:** The Ordinary Portland cement of 53 grades.
- **Fine aggregate:** Locally available sand passing through 4.75mm sieve.
- **Waste Rubber Tyre Crumb** as a partial replacement.
- **Coarse aggregate:** Locally available crushed blue granite stones aggregate of nominal size 10mm.
- **Water:** Casting and curing of specimens are with portable water that is available on site.
- **Concrete Admixture:** For good workability, water reduction, waterproofing, high performance.
- **Binding wire:** 16-gauge wire for tying reinforcement with each other.

Coal Tar Epoxy Paint. Figure 3.

Figure 3: Coal Tar Epoxy Paint

- **Bamboo:** The brown color of 3 years old with largest diameter culm.

9. Waste Rubber Tyre Crumb as a replacement of fine aggregate. Figure 6.

Figure 6: Waste Rubber Tyre Crumb.

- **Mould:** As per the selected size and thickness of the tremix concrete panel.

Equipment Used in research work:

1) Vacuum pump: A vacuum pump is a small but strong pump of 5 to 10 HP. Water is extracted by vacuum and stored in the water separator. The mats are placed over fine filter pads, which prevent the removal of cement with water.

2) Water separator: Proper control on the magnitude of the water removed is equal to the contraction in the total volume of concrete. About 3% reduction in concrete layer depth takes place. The filtering pad consists of a rigid backing sheet, expanded metal, wire gauze, or muslin cloth sheet.

3) Filtering pad: A rubber seal is also fitted around the filtering pad filtering pad should have a minimum dimension of 90cm x 60cm.

4) Screed board vibrator: Vibro screed is used for the leveling as well as compaction of concrete. It consists of a high-quality steel bar (4.2 meters) with a spacing of 250 mm in between. A special water protective vibrator motor is mounted in the center.

5) Power Floater: Power floater is used for smoothness on Bamboo reinforced flooring surfaces.

6) Groove Cutter: The groove cutter machine is also known as concrete cutter.

Discussion Cbout Testing

1) Workability Test: Workability tests were performed using Slump moulds as it is the quick measure of workability of concrete mixes. The slump test was done in accordance with the IS 1199-1959.

2) Initial Setting Time of Cement: Initial setting time is the time period between the time water is added to cement and the time at which 1mm square section needle fails to penetrate the cement paste, placed in the Vicat's mould 5mm to 7mm from the bottom of the mould.

3) Final Setting Time of Cement: Final setting time is the time period between the time water is added to cement and the time at which 1mm needle makes an impression on the cement paste in the mould but 5 mm attachment does not make any impression.

4) Compressive Strength Test: Compressive strength test was performed according to IS 516: 1959. Cubes of a specimen of size 150 mm x 150 mm x 150 mm were prepared for each mix. After 24 hours the specimens were de-molded and cured in water for 7days, 14 days, and 28 days until testing. The values of compressive strength are 65%, 90%, 99% respectively. For specimens with uneven surfaces, capping was used to minimize the effect of stress concentration. The compressive strength reported is the average of three results obtained from three identical cubes.

5) Flexural Strength Test: The bed of the testing machine shall be provided with two steel rollers, 38 mm in diameter, on which the specimen is to be supported, and these rollers shall be so mounted that the distance from center to center is 60 cm for 15.0 cm specimens or 40 cm for 10.0 cm specimens. The load shall be applied through two similar rollers mounted at the third point of the supporting span that is, spaced at 20 or 13.3 cm center to center.

6) Tensile Strength Test: The splitting tests are well-known indirect tests used for determining the tensile strength of concrete. The test consists of applying a compressive line load along with the opposite generators of a concrete cylinder placed with its axis horizontal between the compressive platens. The test was performed according to the procedure adopted in IS 5816: 1999. Cylinders of specimen size 150 mm x 300 mm were prepared for each mix.

DESCRIPTION OF THE DIAGRAM

Fig. 1: Flowchart of Methodology.

Fig. 2: Splitting of Bamboo in Strips.

Fig. 3: Coal Tar Epoxy Paint.

Fig. 4: Treatment on Strips of Bamboo with Epoxy Coal Tar Paint.

Fig. 5: Bamboo Reinforcement.

Fig. 6: Waste Rubber Tyre Crumb.

Fig. 7: Dimensions of Tremix Concrete Panel.

Fig. 8: Tremix Concrete Panel.

Fig. 9: Test Setup for Testing of Tremix Concrete Panel in UTM

BRIEF DESCRIPTION OF THE DIAGRAM

Flowchart of Experimental Work

According to the literature, survey bamboo cannot be used directly as reinforcement in concrete. Thus, bamboo has to be processed to make it compatible and improve its performance in concrete. Hence the process of selection, seasoning, and treatment is done before use. Figure 1 shows the flowchart of methodology for the preparation process of bamboo reinforced tremix concrete flooring .

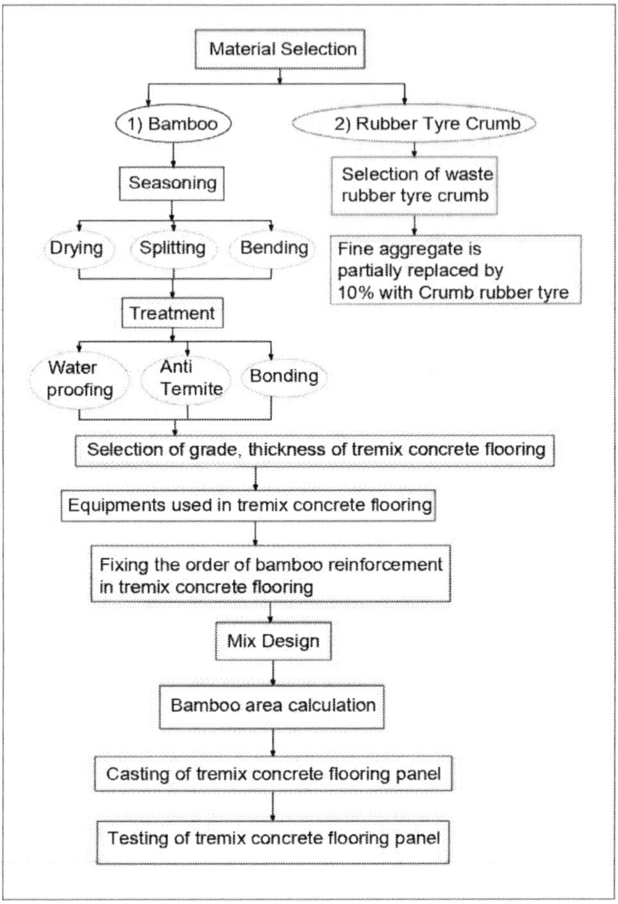

Figure 1: Flowchart of Methodology.

Selection of Bamboo

- Using bamboo as reinforcement in concrete, the following points have to be considered.
- The bamboo appearing brown should be made for use, confirming that the plant is a minimum of three years old.
- Select the longest and largest diameter culms available, which ensure complete growth of culm.
- The complete green, unseasoned bamboo should not be considered for use as these culms are generally weaker due to the high moisture content.

Seasoning of Bamboo

Seasoning is a process that will make bamboo compatible to use as reinforcement in concrete

- **Drying**: the harvested bamboo should be allowed to dry for three to four weeks before use to reduce the moisture content from the fibers
- **Splitting:** the bamboo can be done using a splitting machine or using a sharp knife to obtain straight sectional strips. Figure 2 shows the splitting of bamboo in strips.

Figure 2: Splitting of Bamboo in Strips.

- **Bending:** Bamboo strips can be bent with the application of pressure or heat. These bamboo strips can be used as an outer ring as a circumferential reinforcement to avoid slipping at the edges.

Treatment of Bamboo

To improve the performance of bamboo as reinforcement, the following treatments are to be carried to get optimal results. Figure 4 shows the treatment done on bamboo strips.

Figure 4: Treatment on Strips of Bamboo with epoxy coal tar paint.

- **Waterproofing (Low water absorption):** bamboo should be coated with coal tar epoxy paint for low water absorption as bamboo has the capacity to absorb up to 50% of the water of its weight from concrete.
- **Anti-termite:** coating or soaking in lethal solution to resist termite attack.
- **Good bonding in concrete:** a sprinkling of sand on the coated bamboo with coal tar epoxy paint and dried for 24-48 hours.

Selection of Grade and Thickness of Tremix Concrete Flooring Panel

The Grade of tremix concrete floor selected in this study light duty & heavy-duty square, 600x600x200mm in dimension and was tested for load resisting capacity of tremix concrete flooring panel.

Mix Design

The grade of concrete selected is M30 as per the recommendation in IS 12592:2002, and for bamboo reinforced, the water-cement ratio must be as low as possible because bamboo can absorb water from concrete, and the tensile strength of bamboo may reduce. Table 2 shows the mix design of M30 grade concrete.

Table 2: Mix Design of M30 Grade

M30 Grade	Materials	kg/m^3
1	Cement	360
2	Water	144
3	Fine Aggregate	812
	Sand	751.10
	Waste Rubber Tyre Crumb (10%)	81.20
4	Coarse Aggregate	1284
	20 mm coarse Aggregate	645
	10 mm coarse Aggregate	639
5	Admixture	1.8
6	Water Cement Ratio	0.40

Bamboo Area Calculation (Ratio of Bamboo Reinforcement)

For bamboo as reinforcement in concrete, the ratio varies from 2-4% of the area of cross-section. The bamboo reinforcement ratio is taken as 4% of the area of cross-section of manhole cover.

Manual Design Calculation of Tremix Concrete Flooring

For the design purpose, a tremix concrete floor is considered as a simply supported slab.

Design for Light Duty Bamboo Reinforced Tremix Concrete Floor Panel size 600x600x200 mm (4% Bamboo Reinforcement of the Area of Cross-section) figure 7 shows the dimension of tremix concrete floor panel.

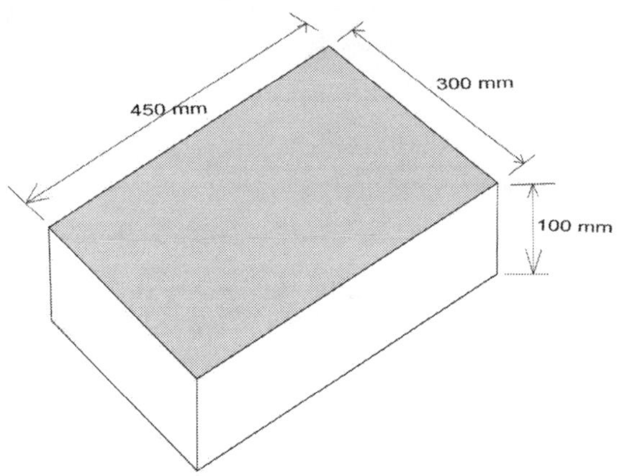

Figure 7: Dimensions of Tremix Concrete Panel.

The grade of concrete taken is M30, and bamboo tensile strength is considered as 125 N/mm2, and bamboo bars are assumed to be square in cross-section as steel bars

Step 1: Type of slab

Ly/Lx = 0.450/0.300 = 1.5< 2 .....(Hence two-way slab)

Ly is the longest span, and Lx is the shortest span of the slab.

Step 2: Depth of slab (IS 456:2000 clause 23.2.1)

d = span / BV * MF = 450 / 20*1.5 = 15mm

d is taken as 85mm for further calculation

D = 100mm

BV is the basic value, MF is modification factor, d is effective depth, and D is total depth

Step 3: Effective span (IS 456:2000 clause 22.2)

Leff = length of slab + effective depth (d) = 450 + 35 = 485 mm

Step 4: Load calculation

Dead load = density of RCC* volume = 25*0.0135= 0.34 KN/m

Live load = target load / length of slab = 25KN / 0.45m = 55.55 KN/m

Total Load = dead load + live load = 55.89 KN/m

Working load (W) =1.5* total load = 1.5 * 55.89 = 83.83 KN/m

Step 5: Ultimate moment (IS 456:2000 clause D-2.1)

αx=0.062, αy=0.062 (IS 456:2000 Table 27)

Mx= $\alpha x * W * Leff^2 = 0.062 * 83.83 * (0.335)^2$ =0.58 KNm

My= $\alpha y * W * Leff^2 = 0.062 * 83.83 * (0.485)^2$ = 1.22 KNm

Md (ultimate moment) = 1.22 KNm

αx and αy are moment coefficient

Step 6: Check for depth

Xu max = 0.46 d ... (Xu max taken is assumed)

Md = 0.36 * fck * Xu max * b (d- 0.42 Xu max)

$1.22 * 10^6 = 0.36 * 30 * 0.46\ d * 1000 * (d - 0.42 * 0.46\ d)$

d = 17.44 < 85 mm

fck is compressive strength of concrete

Step 7: Reinforcement calculation

Abr(long span)= (0.5*fck/fy) * (1-(in squareroot(1-((4.6*Md)/fck. bd^2))*bd

= (0.5*30/125)*(1-(insquareroot(1-((4.6*1.22*10^6)/

30*1000*85^2))*1000*85

Abr (long span)= 132.91 mm ^{2}

Abr(short span) = (0.5*fck/fy) * (1-(in squareroot(1-((4.6*Md)/fck. bd^2))*bd

= (0.5*30/125)*(1-(insquareroot(1-((4.6*0.58*10^6)/

30*1000*85^2))*1000*85

Abr(short span) = 62.97 mm ^{2}

Abr is the area of bamboo reinforcement; fy is the tensile strength of bamboo (125 N/mm^2).

Step 8: Minimum reinforcement

0.12 % * bd = 0.12 / 100 * 1000 * 85 = 102 mm^2

Step 9: Spacing

Spacing = ((πd^2/4) / (Abr required))* 1000 = ((3.14*12^2/4) / (132.91))*1000

= 850.19 mm

Hence use 4 numbers of 12 mm bamboo bars are used in x at 90 mm spacing and 2 numbers of 12 mm bamboo bars are used in y at 90mm spacing center to center.

d is the diameter of the bar.

SUMMARY

The grade of concrete used is M30 grade, and approximately 4% of the area of cross-section as bamboo reinforcement ratio and cross-sectional area of a single bamboo strip are approximately 113 mm^2. 4 number of 12 mm bamboo bars are used in x at 90 mm spacing and 2 number of 12 mm bamboo bars are used in y at 90mm spacing center to center Figure 4 shows bamboo reinforced manhole cover.

The Casting of Tremix Concrete Floor Panel

For the casting process, the mesh of bamboo strips tied with the binding wire at a required spacing and kept inside the mould (formwork), and to both sides of bamboo and then placing concrete with the required amount of tamping is done. Figure 5. shows Bamboo Reinforcement and figure 8. shows the casting of tremix concrete floor panel.

Figure 5: Bamboo Reinforcement

Figure 8: Tremix Concrete Panel.

Testing on Tremix Concrete Floor Panel (Load Resisting Capacity of Tremix Concrete Floor Panel)

Load test on tremix concrete floor panel is done in UTM machine, in this test application of load on the tremix concrete floor panel is done at a rate of 1 KN/s up to its target load with the use of 450mm x 300mm diameter rectangular panel (plate) placed on its geometrical center as per IS code 12592:2002 recommendation. Figure 9. shows the test setup for a testing tremix concrete floor panel in UTM.

Figure 9: Test Setup for Testing of Tremix Concrete Panel in UTM

PROBLEM STATEMENT

No one can assure that an upcoming generation will experience an eco-friendly environment, as we all are facing the major issue of global warming overall the world. In most developing countries, the demand for conventional construction materials leads to the overexploitation of natural resources required for their manufacturing along with their adverse effects on the environment. Increased carbon emission is one of the main reasons for global warming. Steel and cement are widely used materials in the construction industry and are releasing more and more carbon gases into the environment during their manufacturing process. Along with the pollution, one more downside of steel is its day-by-day rising cost. Every year, around a billion scrap tires reach the end of their useful lives around the world. Storage of waste rubber tyre is expensive and requires large areas since tyre storage is made voids.

Disposal of waste rubber tyre is also the major problem, as it makes more land to be occupied and causing a hazard to the society.

WE CLAIM

Our invention gives importance to natural and eco-friendly materials for supporting sustainable development. The non-renewable materials such as steel are being replaced with bamboo which has the potential of getting similar results as steel reinforcement in the concrete according to many studies. The process of selection, seasoning, and treatment of bamboo is carried out before to improve its properties and performance as reinforcement in concrete this research investigates the use of bamboo as reinforcement in the production of reinforced Tremix concrete flooring as an alternative replacement to steel reinforcement. The purpose of this study was to investigate the load resisting capacity of bamboo reinforced tremix concrete flooring in order to apply this design in construction practice. Testing, grade, and shape are considered as per IS |12592:2002|. This research work makes an approach to develop a new material that would contribute to sustainable development and also green alternative material to steel. This material will no longer become an issue in terms of theft and would contribute towards sustainable development. Rubber has a very low modulus of elasticity creating higher internal tensile stress that acts perpendicular to the direction of load. The rubber act as the filler of the pores and it also makes the concrete floor stronger it also helps to reduce shrinkage cracks in concrete flooring. In this way, the bamboo and waste rubber tyre crumb will be the perfect substitute for constructing the modified and sustainable concrete floorings. This will help to reduce air pollution, lower the cost, benefits the farmer, and generate the source of employments in various ways.

- According to claim1# the invention is to the behavior of bamboo reinforcement is similar to steel reinforcement, and the application of bamboo reinforcement in tremix concrete floorings will achieve the desired strength of light-duty tremix concrete flooring of |+25KN|.
- According to claim 1# the rubber act as the filler of the pores and it also makes the concrete floor stronger it also helps to reduce shrinkage cracks in concrete flooring.
- According to claim1, 2# the invention is to the bamboo reinforcement, the durability of tremix concrete flooring will increase as it overcomes the problem of corrosion of steel reinforcement.

- According to claim1, 2, 3# the invention is to the Bamboo reinforcement in tremix concrete floorings, the cost can be reduced as compared to steel reinforced tremix concrete flooring.
- According to claim1, 2, 4# the invention is to the carbon footprint emission can be reduced using bamboo reinforcement in tremix concrete floorings, which can lower the effect of global warming and contribute towards sustainable development.

CITATIONS

- 202221002500, Dr. Biplab Kumar Sarkar, Designing the New Generation of Filtration Devices. MAHATMA EDUCATION SOCIETY'S, PILLAI HOC COLLEGE OF ENGINEERING &TECHNOLOGY RASAYANI, TALUKA PANVEL,
- 202211002503, Dr. Biplab Kumar Sarkar, Growing Your Small Business Faster Management System MAHATMA EDUCATION SOCIETY'S, PILLAI HOC COLLEGE OF ENGINEERING &TECHNOLOGY RASAYANI, TALUKA PANVEL,
- 202211002504, Dr. Biplab Kumar Sarkar, Planning and Forecasting in Specific Business Organization Management System, MAHATMA EDUCATION SOCIETY'S, PILLAI HOC COLLEGE OF ENGINEERING &TECHNOLOGY RASAYANI TALUKA PANVEL
- 202221002497, Dr. Biplab Kumar Sarkar, Method and Process Quick Electric Vehicle Charging. MAHATMA EDUCATION SOCIETY'S, PILLAI HOC COLLEGE OF ENGINEERING &TECHNOLOGY RASAYANI, TALUKA PANVEL
- 202221001659, Dr. Biplab Kumar Sarkar, Affordable Cost High-Performance VLSI Architecture for Montgomery Modular Advanced Multiplication, MAHATMA EDUCATION SOCIETY'S, PILLAI HOC COLLEGE OF ENGINEERING &TECHNOLOGY RASAYANI, TALUKA PANVEL
- 202221001660, Dr. Biplab Kumar Sarkar, Investigating the Cases of Novel Coronavirus Disease (Covid-19) Using Intelligent Dynamic Statistical Techniques, MAHATMA EDUCATION SOCIETY'S, PILLAI HOC COLLEGE OF ENGINEERING &TECHNOLOGY RASAYANI, TALUKA PANVEL,
- Dr. Biplab Kumar Sarkar, 202241000869, MAHATMA EDUCATION SOCIETY'S PILLAI HOC COLLEGE OF ENGINEERING &TECHNOLOGY RASAYANI TALUKA PANVEL," Unique Plug-In Hybrid Electric Vehicles"
- Dr. Biplab Kumar Sarkar, 20223221736270ODE, MAHATMA EDUCATION SOCIETY'S, PILLAI HOC COLLEGE OF ENGINEERING &TECHNOLOGY RASAYANI TALUKA PANVEL, "A DIGITAL VOTING AND FRAUD DETECTION SYSTEM USING BLOCK CHAIN AND MACHINE LEARNING" 2022, Indian Patent Office.
- Dr. Biplab Kumar Sarkar, 20221011560, MAHATMA EDUCATION SOCIETY'S PILLAI HOC COLLEGE OF ENGINEERING &TECHNOLOGY RASAYANI TALUKA PANVEL, "Advanced Method and Process Quick Electric Vehicle Charging." 2022, Indian Patent Office.

- Dr. Biplab Kumar Sarkar, 202241011563, PILLAI HOC COLLEGE OF ENGINEERING & TECHNOLOGY RASAYANI TALUKA PANVEL, DIST, NAVI MUMBAI, MAHARASHTRA 410207, INDIA." Cell Capacity Evaluation using Fifth Generation WLAN and IEEE 802.11ax for VBR Traffic." 2022, Indian Patent Office.
- Mahatma Education Society's Pillai College of Architecture, Ar. Shubhangi Bhide, 202221011555, Mahatma Education Society's Pillai College of Architecture, Plot. No.10, KM Vasudevan Campus, Sector 16, Panvel, Mumbai, "Design of footrest &Vertical Support for Indian Vehicle." 2022, Indian Patent Office
- Dr. Shabab Ahmed Rizvi, 202221012727, M.A (Economics), Mphil (Economics) M.B.A (HR) Ph.D, Assistant Professor, MAHATMA EDUCATION SOCIETY'S, Pillai College of Arts Commerce & Science. New Panvel, India." Self-charging Battery Bank" 2022, Indian Patent Office.
- Skeeriti Asset LLP,Mr Nikhil D. Hadap, Dr. Arti Hadap, Dr. Tusharika Banerjee, 202221016225.," Dry – Storage – Cabinet" Faculty at EXTC Department, Pillai Coellege of Engineering, 2022, Indian Patent Office.
- Dr. Biplab Kumar Sarkar, 202221018017, PILLAI HOC COLLEGE OF ENGINEERING &TECHNOLOGY RASAYANI TALUKA PANVEL, "Vibration of a single walled carbon Nano-tube for delivering of Nanoparticles." 2022, Indian Patent Office.

Chapter - 10

Manufacturing of Lightweight Sandwich Bricks Using a Unique Combination of Bamboo Fiber and E-Waste Powder

Rohit Devappa Gawade, Prof. (Dr.) Shilpa Pankaj Kewate

Mahatma Education Society's, Pillai Hoc College pf Engineering &Technology Rasayani
Taluka Panvel, Dist, Navi Mumbai, Maharashtra 410207, India

ABSTRACT

Bricks are widely used as a building material in construction industry around the world. For sustainable growth of construction industry there is need of utilizing alternative materials in construction and utilization of waste material. Management and Utilization of various types of wastes like E-waste, Hazardous waste, Agricultural waste, Industrial waste, Municipal waste, Commercial waste, Chemical waste, etc is a major proplem socciety is facing now a days. This reserch work proposes manufcturing of Bricks from from natural waste material (coconut and bamboofibre) & industrial waste (E-waste powder and fly ash). The table tennis balls are used to reduce density of bricks and to reduce the use of cement.This newly developed brick will have low density and hence brick will be light in weight. This newly developed product will be Eco-friendly and more sustainable.Comparitive study betwwen tradional brick and newly developed brick will be studied based on various test on bricks i.e. Compressive Strength, Water Absorption,Soundness Testing Corrosion Testing Hardness Test

FIELD OF THE INVENTION

Light Weight Sandwich Brick is combination of Bamboo Fibre and E-water Powder.

BACKGROUND OF THE INVENTION

Clay brick remains one of the most crucial building materials in India. Clay brick is the conventional Industry in India, usually limited to rural areas. Mostly, the Indian brick industry, with more than One Lakh manufacture units producing Two Hundred Billion bricks annually. It is the 2nd leading brick producer in the world after China. The industry has a yearly turnover of Rs.10,000cr, and very importantly, it is one of the major employment producing industries, employing millions of labours. In Uttar Pradesh Bricks are made with waste plastic material to reduce pollution in that state with growing nowadays.

Fly ash, Coconut Fibre, Bamboo Fibre, E-waste Powder is natural eco-friendly material. Compressive strength in the brick can be achieved by Fly ash, Coconut fibre, Bamboo fibre, E-waste powder. In this brick density of brick will reduce and have a brick is in Light Weight so that anyone can move safely and easily from one place to another, this brick is 100% more Eco-friendly and more sustainable. And main importance is we are achieving Low density of brick.

The OPC-53 grade of cement of Lightweight brick and eco-friendly can achieve strength. Which can be used in residential buildings, pavements and more. The shape of Brick is Rectangular (Cuboidal) with dimensions 190 x 90 x 90mm. The eco-friendly bricks ratio depends on the volume of bricks. The grade of cement used is OPC-53 grade with the addition of 1% of plasticizer and Percentage of all these ingredients like Fly ash, Coconut Bamboo Fibres and E-waste Powder of its volume of brick.

The product developed would be sellable in the market. From the literature survey, it has been observed that the availability of Brick has become a major issue related to the cost and weight of the brick. Fibres are the naturally available resource. It can be used to replace cement and sand. Author's target market competition would with established Eco-friendly LW Brick producing companies and target market would be like Brick Industries, Manufacture and suppliers, and Government Brick company like "Brick India Pvt Ltd" Most online Company like "India Mart".

BAMBOOFIBRE

Bamboo is a remarkable and highly versatile natural resource. For many centuries it's has been utilized in Asia during a range of uses, including cooking, construction, transport, textiles and medicine. The Western world has only relatively recently discovered that bamboo offers a good range of advantages and an eco-friendly solution to several of our modern requirements.

There are a few more benefits of bamboo fibre: it allows the material to be extremely soft, durable, strong, absorbent, lightweight, and dry quickly.

Advantages of BambooFibre

Bamboo is a natural, renewable, and biodegradable material. Its tensile strength to specific weight ratio is high. Also, it possesses both tensile and compressive strength. Therefore, the cost of bamboo is cheaper as its availability is abundant in rural and urban regions. In addition, it can prevent pollution by absorbing a large amount of nitrogen from wastewater.

Disadvantages of BambooFibre

The chemicals wont to process the material are harmful to human health and therefore the environment. Shrinkage of the Fabric: Bamboo fabric tends to shrink at a faster rate compared to cotton. Expensive: Natural bamboo fabric tends to be costlier than the rayon type or maybe cotton.

BALL USED IN BRICK

The small table tennis ball is going to be used in this brick. The small ball of table tennis hasa diameter of 40 mm = 4cm. By calculation of standard brick and ball, we conclude that we r using 2 or 3 balls in the brick.

- Benefits of Ball used in brick.
- By using a ball in brick, the requirement of cement per brick will be less.
- By using a ball hollow part that will occur inside the brick means cement will be cover the outer surface of the ball.
- By using a ball in brick Density of brick will be reduced compared to other standard clay brick.

OBJECTIVES OF THE INVENTION

The following are the objectives that will be studied in the research work:

1. To explore the use of Coconut fibre, Bamboofibre and E-waste powder as binding material in brick.
2. To carry out an experimental investigation on Coconut fibre, Bamboofibre and E-waste powder as binding material in brick.
3. To compare this newly developed product (lightweight sandwich bricks) with conventional brick with the following parameter:
    - Compressive Strength.
    - Cost
    - Temperature Testing
    - Water Absorption
    - Soundness Testing
    - Corrosion Testing
    - Hardness Test
    - Structure Test
    - Shape Size Color Test
    - Mortar Bonding Test

SUMMARY OF THE INVENTION

Materials Used in research work:

- **Cement:** The Ordinary Portland cement of 53 grades.
- **Water:** Casting and curing of specimens are with portable water that is available on site.
- **Fly-ash:** Class F: Fly ash normally produced by burning anthracite or bituminous coal, usually has less than 5% CaO. Class F fly ash has pozzolanic properties only.
- **Class C:** Fly ash is normally produced by burning lignite or sub-bituminous coal. Some class C fly ash may have CaO content above 10%. In addition to pozzolanic properties, class C fly ash also possesses cementitious properties. Fly ash used is of type class C with a specific gravity of 2.19.

Fly-ash material is shown in fig 3.

Figure 3: Fly-ash Powder.

- **BambooFibre:** The Bamboo Fibreis shown in fig 4.

Figure 4: Bamboo Fibre

- **Coconut Fibres:** Coconutfibresare shown in Figure 5.

Figure 5: Coconut Fibres.

- **E-waste Powder:** The electronic waste powder used in brick has fine size. E-waste powder is shown in fig 6.

Figure 6: E-waste Powder.

- **Table Tennis ball:** The table tennis balls are used to reduce density of bricks and to reduce the use of cement.

Ball is shown in fig 8.

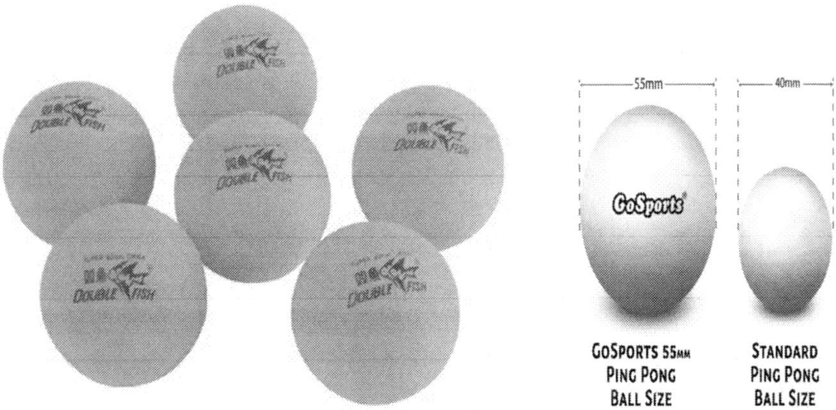

Figure 8: Table Tennis Balland Size

- **Mould:** As per the selected shape and size of the standard brick 190 x 90x 90 mm.

The figure of the brick is shown in fig 7.

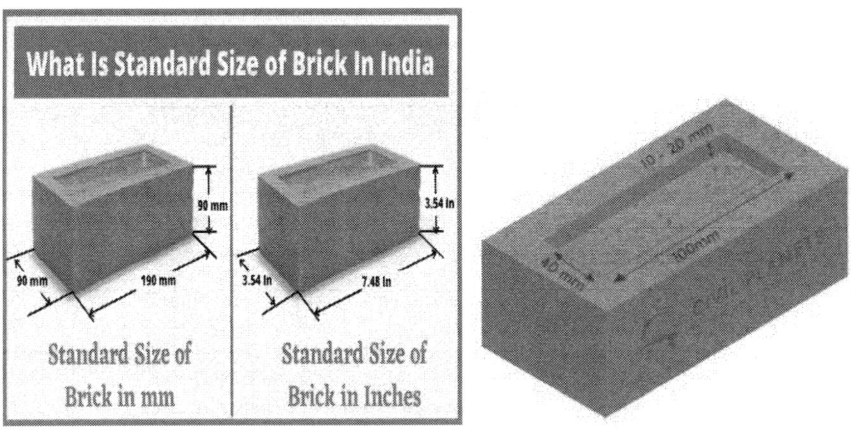

Figure 7: Dimensions of Standard Brick.

Mould cast brick with a cross-section of brick and ball fig shown in fig 9.

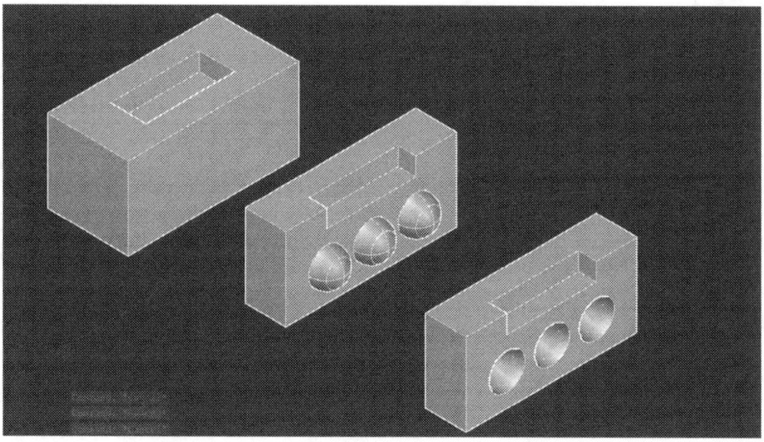

Figure 9: Cross section of brick (A-B).

- **Superplasticizer:** Superplasticizers are used as Dr Fixit in Brick.

DESCRIPTION OF THE DIAGRAM

Fig. 1: Flowchart of Methodology.

Fig. 2: Ordinary Portland Cement of 53 Grade.

Fig. 3: Flyash

Fig. 4: Bamboo Fibre.

Fig. 5: Coconut Fibre

Fig. 6: E-waste Powder

Fig. 7: Dimensions of Standard Brick

Fig. 8: Table tennis ball and size

Fig. 9: Cross-section of brick.

Fig. 10: Test Setup for Brick in UTM.

Fig. 11: Superplasticizer.

BRIEF DESCRIPTION OF THE DIAGRAM

Flowchart of Methodology

According to the literature, survey the Combination of Fly ash, Coconut Fibre, E-waste powder and Bamboo Fibre had not been used till now in the brick to achieve strength and less density. Thus, bamboo Fibrehas to be processed to make it compatible and improve its performance in brick. The flowchart showing the methodology for the preparation process of a Lightweight brick is shown in Figure 1.

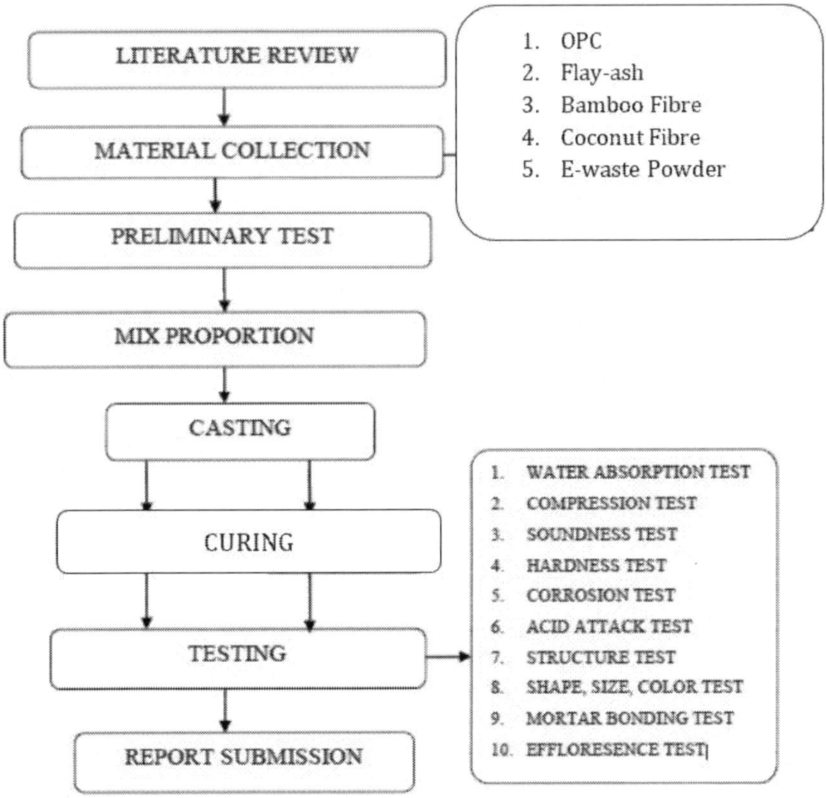

Figure 1: Flowchart of Methodology.

Ordinary Portland Cement (53 Grade)

OPC 53 grade cement is generally used for **non-structural works** like plastering flooring etc. Whereas OPC 53 grade cement is employed within the projects which require higher strength like concrete bridges, runways RCC works etc.

53-grade cement is an **Ordinary Portland Cement (OPC)or Standard Hydraulic Cement**, which is that the most generally used cement in India. 53-grade cement attains 27 MPa (compression strength) in 3 Days. It is also the fastest setting cement among other grades.

Bamboo Fibre

Using bamboo as fibreinbrick, the following points have to be considered.

- Bamboo fibremay be a regenerated cellulosic fibre produced from bamboo. Starchy pulp is produced from bamboo stems and leaves through a process of alkaline hydrolysis and multi-phase bleaching. Further chemical processes produce bamboo fibre.
- Repeated technological analysis has proved that type of fibre has a thinness degree and whiteness degree close to normal finely bleached viscose and has strong durability, stability and tenacity.
- Bamboo Fibrefabric is formed of 100% bamboo pulp fibre. It is characterized by its good hygroscopicity, excellent permeability, soft feel, easiness to straighten and dye and splendid colour effect of pigmentation.

Selection of Grade, Shape and Size of Material Required for the Casting of Brick.

Fly-Ash

The Grade of Fly-ash isselected in between F class and C class. **Class F fly ashes** are generally more effective than Class C fly ashes in reducing the heat of hydration. Not only is the risk of thermal cracking reduced, but greater ultimate strength is attained in concrete with fly ash because of the pozzolanic reaction.

Coconut Fibre

Fibre will be achieved from the waste material of coconut cover. Fibre is cut into pieces of size 10mm.

Cement

There are 3 types of grades of cement 33 grade, 43 grade and 53 grade. 53 Grade of cement is being used in this lightweight brick because 53 grade gives more strength to the material.

E-waste Powder

E-waste means electronic waste which is discarded waste. In this brick, we use discarded electronic waste and we crushed the material into powder form in micron form.

Superplasticizer

Superplasticizer is being used in this brick to achieve strength good workability. Plasticizers of Dr, Fixit is being used in this brick.Figure 11 shows Superplasticizer

Figure 11: Superplasticizer

Mix Design

The grade of cement selected is 53 grade, and for bamboo fibre and all fibre used in this brick, the water-cement ratio must be as low as possible because bamboo fibrecan absorb water from cement. Table 1 shows the mix design of 53-grade cement.

Table 1: Mix Design of 53 Grade

M30 Grade	Materials	Trial 1 (%)	Trial 2 (%)	Trial 3 (%)
1	Cement	45	48	50
2	Water			
3	Fly-ash	25	28	30
4	Coconut Fibre	10	12	10
5	Bamboo Fibre	5	8	8
6	E-waste Powder	2	1.5	1
7	Superplasticizer	2	1.5	1
8	Water Cement Ratio	0.5	0.4	0.5

Manual Design Calculation of Light Weight Sandwich Brick

Casting of Brick

For the casting process, the cement, bamboo fibre, coconut fibre, E-waste powder and Ball is placed inside the brick used in brick casting and kept inside the mould (formwork)and then placedcement with the required amount of tamping is done.

Testing on Brick

Load test on Brick is done in UTM machine, in this test application of load on theBrick is done at a rate of 14 KN/s up to its target load with the use of 190 x 90 mm rectangular top surface placed on its geometrical centre. Figure 10 shows the test setup for a testing brick in UTM.

Figure 10: Test Setup for Testing of Brick in UTM

PROBLEM STATEMENT

A significant problem faced in the world is the issue of global warming.

Coconut fibre is the most well-known fibrous waste from coconut's cultivation. Every year the world produces at least 30 million tons of coconut, which are abundant in coastal areas of tropical countries. The coconut husk is consist of 30% fibre and 70% pith, with high lignin and phenolic content.

E-waste contains a laundry list of chemicals that are harmful to people and therefore the environment, like mercury, lead, beryllium, brominated flame retardants, and cadmium, i.e., stuff that sounds as bad as it is. When electronics are mishandled during disposal, these chemicals end up in our soil, water, and air.

The effect of ash on the environment can vary supported on the thermal power station where it's is produced, also because the proportion of fly ash to bottom ash in the waste product.Fly ash dust can be deposited on topsoil increasing the pH and affecting the plants and animals with in the surrounding ecosystem.

Inhalation or ingestion of the toxins in fly ash can have impacts on the nervous system, causing cognitive defects, developmental delays, and behavioural problems while also increasing a person's chance of developing lung disease, kidney disease, and renal disorder.

From the above statement, we conclude that E-waste is more toxic so we decided to use E-waste powder in this brick. All fibre like bamboo, coconut fibre E-waste crushed powder material in brick to make eco-friendly lightweight brick.

WE CLAIM

In this research work these are the things like E-waste powder, Coconut fibre, Bamboo fibre, fly ash is used and it will help protect the environment. There are wastes like E-waste, Hazardous waste, Agricultural waste, Industrial waste, Municipal waste, Commercial waste, Chemical waste, etc. Bricks are widely used construction and artifact round the world. Bricks are prepared from natural waste material (coconut and bamboofibre) & industrial waste (E-waste powder). In this research work, Bamboo fibre and Coconut fibreare shredded into tiny pieces to enhance the strength of bricks.To study the use of e-waste powder in the manufacturing of brick. Effect of brick on compressive strength, water content and density are studied.To study the impact of partial replacement of cement by E-waste powder and fibre on the properties of concrete, experiments were conducted on different concrete mixes.Soil is employed as a binding material for natural waste material. The main objective of the present study is to reduce the quantity of soil with natural waste material as well as industrial waste material. The coconut fibre, bamboo fibre, fly-ash and e-waste powder are used to make lightweight bricks.

- According to claim1# the invention is to the behaviour of fibre used in research is similar to reinforcement, and the application of fibres used in brick will achieve the desired strength of lightweight sandwich brick.

- According to claim1, 2# the invention is to the fibre reinforcement, the durability of bricks will increase as it overcomes the problem of voids present in the brick.
- According to claim1, 2, 3# the invention is to the fibre reinforcement in bricks the cost can be reduced compared to clay brick.

The non-renewable materials such as sand are being replaced with all fibre like coconut fibre, bamboo fibre, E-waste powder which has the potential of getting similar results as clay brick in the brick according to many studies. The main issue is with the cost and weight of the brick. The purpose of this study was to investigate the load resisting capacity of Ecofriendly Brick to apply this design in construction practice. This research work approaches to develop a new material that would contribute to sustainable development and also green alternative material.

CITATIONS

- Dr. Jagdish W. Bakal, Dr. Biplab Kumar Sarkar, Dr. Reena Singh, 202221018015, MAHATMA EDUCATION SOCIETY'S PILLAI HOC COLLEGE OF ENGINEERING &TECHNOLOGY RASAYANI TALUKA PANVEL, "Genetic level advanced therapy for neonatal gene Treatment" 2022, Indian Patent Office
- Dr. Jagdish W. Bakal, Dr. Lata Menon, Dr. Biplab Kumar Sarkar, Dr. G.R. Patil, Mr. Raju Narwade, 202221018014, MAHATMA EDUCATION SOCIETY'S PILLAI HOC COLLEGE OF ENGINEERING &TECHNOLOGY RASAYANI TALUKA PANVEL," AI- Based Product Ranking Evaluation Process System" 2022, Indian Patent Office.
- Dr. Jagdish W. Bakal, Dr. Biplab Kumar Sarkar, Dr. Reena Singh, Ms. Rohini Bhosale, Ms. Rajashree Gadhave 202221018016, MAHATMA EDUCATION SOCIETY'S PILLAI HOC COLLEGE OF ENGINEERING &TECHNOLOGY RASAYANI TALUKA PANVEL, DIST, NAVI MUMBAI, MAHARASHTRA 410207, INDIA," Point Regression Analysis-Based Big Data Processing Using Machine Learning and Deep Learning" Indian Patent Office.
- MES's Pillai College of Education and Research Chembur, 202021106864, Granted, Dr. Reni Francis of MES's Pillai College of Education and Research, Chembur.," Effectiveness of STEAM education in fostering Integrative learning through iLEARN Model" ,Germany Patent office.
- Dr. Daphne Pillai, 202022100827, Granted, Director, Pillai College of Arts, Commerce and Science MES's Pillai College of Arts, Commerce and Science, Dr. K. M. Vasudevan Pillai Campus, Sector 16, New Panvel, 401206 Navi Mumbai, India," New methods of education to meet post-Covid requirements" Germany Patent Office.
- Dr. Biplab Kumar Sarkar, Dr. Lata Menon, US 2022/0108803, MAHATMA EDUCATION SOCIETY'S PILLAI HOC COLLEGE OF ENGINEERING &TECHNOLOGY RASAYANI TALUKA PANVEL, "Method for covid-19 Detection to spred" ,US-Patent.
- Meera Ranjith, Neelam Raju Gupta, Gnana Vidya, Jasprit Kaur Kawaljit Singh Gupta, Poorni Raju Swamy, Dr. B.K. Kumar ,GE-18-1-32," MAHATMA JUNIOR COLLEGE OF EDUCATION ,OPPOSITE FIRE BRIGADE,SION TROMBAY ROAD,CHEMBUR NAKA,MUMBAI ," Teaching Kitchens as learning labs for life skills in secondary school curriculum" 2022, Indian Patent Office.
- Dr. Biplab Kumar Sarkar, 202241920224, Dr. Biplab Kumar Sarkar, MAHATMA EDUCATION SOCIETY'S PILLAI HOC COLLEGE OF ENGINEERING &TECHNOLOGY

RASAYANI TALUKA PANVEL, "Mobile Based Application for Automatic Irrigation and Fertigation." , 2022, Indian Patent Office,

- Dr. G.R Patil, Dr. Shilpa Pankaj Kewate, Mr. Raju Narwade , 202221020211, MAHATMA EDUCATION SOCIETY'S PILLAI HOC COLLEGE OF ENGINEERING &TECHNOLOGY RASAYANI, TALUKA PANVEL, DIST, NAVI MUMBAI, MAHARASHTRA 410207, INDIA," Mobile Based Application for Automatic Irrigation, Fertigation and Notification" 2022, Indian Patent Office.
- Vaishnavi Veeraraghavan, Dharini Veeraraghavan, Dr. Biplab Kumar Sarkar, 202221021776, MAHATMA JUNIOR COLLEGE OF EDUCATION, OPPOSITE FIRE BRIGADE, SION TROMBAY ROAD, CHEMBUR-400071," Innovation for the Visually Challenged" 2022, Indian Patent Office,
- Rohini Bhosale, 202241021998, MAHATMA EDUCATION SOCIETY'S .PILLAI HOC COLLEGE OF ENGINEERING &TECHNOLOGY RASAYANI,TALUKA PANVEL, "IoT-Based Automatic Railway Gate Controller and Notify." 2022, Indian Patent Office,
- Ankit Arun Ulvekar, Juily Dinesh Nimbolkar, Rushikesh Suhas Haladkar, Kunal Sudhakar Gaikwad, Ms. Manisha B. Jamgade, Dr. G R Patil, 202221021777, MAHATMA EDUCATION SOCIETY'S, PILLAI HOC COLLEGE OF ENGINEERING &TECHNOLOGY RASAYAN PANVEL, "Recycled Concrete Using Demolition Waste with Rice Husk and Coir fibres For Sustainability" 2022, Indian Patent Office.
- 202221023844, Ar. Shubhangi Bhide, Mahatma Education Society's Pillai College of Architecture, "ECO FRIENDLY IV & Saline Stand OF Bamboo Material" 2022, Indian Patent Office.
- Mimi Cherian, Satishkumar L. Varma , 202221021778, Mahatma Education Society's, " SDN Enabled Secure Framework for IoT using Dynamic Counter-Based Approach and Deep Learning Model", 2022, Indian Patent Office.
- Ashish Purushottam Kulkarni, Purushottam Gopalrao Kulkarni, 202221024947, MAHATMA EDUCATION SOCIETY'S, PILLAI HOC COLLEGE OF ENGINEERING &TECHNOLOGY RASAYANI, TALUKA PANVEL, DIST, NAVI MUMBAI," Minimization of accidents due to the Extra gap between the Edge of Platform and Footboard of the Train." 2022, Indian Patent Office.
- Meera Ranjith, Neelam Raju Gupta, Gnana Vidya, Jasprit Kaur Kawaljit Singh Gupta, Poorni Raju Swamy Dr. B.K. Sarkar ,202022102061, Granted, Mahatma Junior College of Education ,Opposite Fire Brigade,Sion Trombay Road, Chembur Naka,Mumbai 400071, India, " Teaching Kitchens as learning labs for life skills in secondary school curriculum" Germany Govt
- Dr. B.K .Sarkar ,202022101506, Granted, MAHATMA EDUCATION SOCIETY'S, PILLAI HOC COLLEGE OF ENGINEERING &TECHNOLOGY RASAYANI, TALUKA PANVEL, DIST, NAVI MUMBAI, MAHARASHTRA 410207, INDIA. "Digital voting and fraud Detection using Block Chain and Machine Learning System." Germany Govt.

- MR.SATPUTE SUNNY SHIVAJI, MR.RANJEET KUMAR YADAV, MR.PRITAM AMBO MUNDHE, PROF.ASOKAN SELVARAJ 202221025962 ,MAHATMA EDUCATION SOCIETY'S, PILLAI HOC COLLEGE OF ENGINEERING &TECHNOLOGY RASAYANI, TALUKA PANVEL, "INTELLIGENT SOLAR HYBRID DEHYDRATOR" 2022, Indian Patent Office
- Mrs. Jyoti Raju Kadam, Mr. Raju A Kadam, Sanjana Raju Kadam202221027979, MAHATMA EDUCATION SOCIETY'S, Assistant Professor, Department Biotechnology, Pillai College of Arts Commerce and Science. (Autonomous), "Intelligent LED Flat Screen Monitor take a Selfies Pictures and Video Recording" 2022, Indian Patent Office.
- Mrs. Jyoti Raju Kadam, Mr. Raju A Kadam, Sanjana Raju Kadam, 202221027981, MAHATMA EDUCATION SOCIETY'S, Assistant Professor, Department Biotechnology, Pillai College of Arts Commerce and Science. (Autonomous) , "Smartphones Convert in to a Mini Projector Lens of High Definition" 2022, Indian Patent Office.
- Mrs. Jyoti Raju Kadam, Mr. Raju A Kadam, 202221027980, MAHATMA EDUCATION SOCIETY'S, 2022, Indian Patent Office. Department Biotechnology, Pillai College of Arts Commerce and Science. (Autonomous), "Built-in Mini Fan of 1 Inch Diameter under Smartphones" 2022, Indian Patent Office.
- Dr. Farhat Fatma Mumtaz Husain Shaikh, Sunita Charanjit Saini, Nikhat Fatma Mumtaz Husain Shaikh 202221028002, MAHATMA EDUCATION SOCIETY'S , Chembur Mumbai, A 302, Neelgagan CHS, Plot No. 33, Sector 42, Seawoods , Nerul West, Navi Mumbai 400706, " Slide Lipstick Holder with Button." 2022, Indian Patent Office.
- Mahatma Education Society's, Prof. Dr. Biplab Kumar Sarkar, 202211028005, Mahatma Education Society's Pillai HOC College Of Engineering &Technology Rasayani, Taluka Panvel, Dist, Navi Mumbai, Maharashtra 410207, India. "-UV-Light Box: Modified Version of UV- Sanitizers", 2022, Indian Patent Office.
- Mahatma Education Society, 202241028431, Mahatma Education Society, Chembur Naka, Mumbai - 400 071, Maharashtra, India." Decresed the Search engine Time Cloud Data based using Machine Lerning" 2022, Indian Patent Office.

Chapter - 11

Precast Prefab Bamboo Reinforced Cement Concrete Compound Wall Panels

Dr. Shilpa Kewate, Mr. Pankaj Kewate

Mahatma Education Society's, Pillai Hoc College of Engineering & Technology Rasayanitaluka Panvel, Dist, Navi Mumbai, Maharashtra 410207, India

ABSTRACT

In earlier days bricks were considered to be the best choice for compound walls to easily work with them. Traditional compound walls are made from clay and their durability depends upon the quality of plaster used while preparation. In terms of material, these walls are very expensive. Traditional compound wall construction involves extensive installation and a hard labor process. To conquer this problem precast compound walls are gaining much attention for a better design. These Precast compound walls are cost-effective and provide long-lasting life. But in the manufacturing process of Precast Reinforced cement concrete (RCC) compound walls, steel is used as reinforcement. The construction sector contributes an extensive amount of carbon footprint from the manufacturing process of steel and cement. Also, due to abundant non-renewable resources used for steel manufacturing, steel prices are rising day by day. Some issues commonly faced with steel are increment in overall cost, use of the non-renewable resources for its manufacturing, polluting environment due to industrial processing, a significant contributor for the overall dead load of structures, etc. So there is a need to have an alternative and environmentally friendly material in construction that won't harm the environment, similar to steel, which can contribute to sustainable development.

Bamboo is natural, renewable, and its tensile strength to specific weight ratio is high. Also, it possesses both tensile and compressive strength. This research

work proposes the use of bamboo as an alternative to steel reinforcement in the manufacturing process of Precast Reinforced cement concrete (RCC) compound walls. This Newly developed product will echo friendly and sustainable as instead of steel Bamboo will be used as reinforcement. At the same time, it will have all the advantages of precast compound walls. This newly developed product will be cost-effective, reusable, lightweight, and durable, save the time of installation and it will have similar strength as that Precast RCC compound wall. This research work proposes eco-friendly and sustainable way towards production of Compound wall.

FIELD OF THE INVENTION

Our Invention is related to precast prefabricated bamboo reinforced cement concrete compound walls. This invention will contribute towards eco-friendly and sustainable way towards production of Compound wall.

BACKGROUND OF THE INVENTION

Manufacture of conventional building materials such as steel, cement, and bricks involves high energy consumption as well as production of greenhouse gases which damages the environment. The average temperature universal has reached alarming levels causing the glaciers to melt and increase in the sea level. Due to which various coastal regions are on the verge of drowning underwater. Countries such as the Maldives have already lost a large chunk of their land due to global warming [2]. Thus there is a great need to look out for alternative materials for the construction industry which should be cheap, sustainable, environmentally friendly, and must have satisfactory structural properties.

Several studies have been undertaken to look out for such unconventional building materials. Literature reveals that a number of materials such as kevlar, polyester, carbon fibers, metal alloys were looked upon as building materials [4]. However, the production of such materials is a difficult process that cannot be manufactured at a village level from the locally available materials and technologies. To resolve the problem, it is essential to develop building technologies that can be used at a village level from the locally available materials. The use of precast prefab Bamboo Reinforced cement concrete technology is one of the best solutions for a sustainable speedy construction approach.

In the present study, a new approach to develop precast prefab Bamboo Reinforced cement concrete compound walls panels has been presented by the authors. This newly developed product is prefabricated, cheap, made from sustainable materials such as bamboo and fly ash, and has great potential in the construction industry.

Bamboo

Bamboo is a part of the grass family of Poaceae, which belongs to the sub-family of Bambusoideae. It is a combined element made up of lengthy and parallel cellulose fibers compressed in a ligneous matrix form (Ghavami, 2004). There is an abundant number of species of bamboo found in the whole world. The carbon sink by 5-year-old bamboo every year is 18 to 23 tons per acre.

Advantages of Bamboo

Bamboo is a natural, renewable, and biodegradable material. Its tensile strength to specific weight ratio is high. Also, it possesses both tensile and compressive strength (Nayak et al, 2013, p. 50). Therefore, the cost of bamboo is cheaper as its availability is abundant in rural and urban regions. In addition, it can prevent pollution by absorbing a large amount of nitrogen from wastewater (Priyadarshee, Chhotu, and Kumar, 2014, p. 14).

Disadvantages of Bamboo

The bamboo tensile strength is relatively much lesser than steel (Khan, 2014, p. 126). Its water absorption capacity is very high, affecting the dimension and developing low bond strength with concrete (Bhonde et al, 2014, p. 128). The structure of bamboo is irregular; thus, the strength may vary drastically. It is a woody grass that can be attacked by termite and decay (Ahmad, Raza and Gupta, 2014, p. 163).

Precast RCC (Reinforced Cement Concrete)

Precast concrete is a construction product produced by casting concrete in a reusable mould which is then cured in a controlled environment, transported to the construction site.

Advantages of Precast concrete

- **Saves Construction Time**: Precast Concrete construction saves time, the risk of project delay is also less.
- **Quality Assurance**: The major factors which regulate the quality of construction such as curing, temperature, mix

design, formwork, etc. can be monitored for Precast Concrete. Therefore, improved quality construction can be performed.

- **Usage of Prestressed Concrete**: By using precast, structural materials of high strength and load-bearing capacity can be achieved, which can result in greater clear span, reduced size of the cross-section of structural members, etc.
- **Cost-effective**: The simplified construction process reduces time, increases productivity, quality and safety and thus the cost is reduced.
- **Durability**: Precast Concrete structure has a longer service time period and minimal maintenance.
- **Aesthetics**: A wide range of shapes and sizes are available to choose from with smooth finishing and thus the aesthetical value of products is increased.

OBJECTIVES OF THE INVENTION

- To provide the society a green product, this has low carbon emission with the use of bamboo reinforcement instead of steel reinforcement.
- To provide the building sector with a newly developed product (Precast prefabricated Bamboo Reinforced cement concrete compound walls) type of compound wall panels that will allow them to maximise their output at a low cost.
- To expedite construction activity by using a 'precast prefab Bamboo Reinforced cement concrete compound walls', thus saving the construction time.
- To deliver the construction industry with a highly reliable compound wall passes bamboo reinforcements
- To replace traditional precast RCC concrete compound walls with 'precast prefab Bamboo Reinforced cement concrete compound walls',

SUMMARY OF THE INVENTION

Materials Used in research work

- **Reinforcement:** Seasoned bamboo the brown colour of 3 years old with largest diameter culm.
- **Bamboo waterproofing agent:** Coal tar liquid for bamboo, epoxy paint

- **Cement:** OPC cement of 53 grade.
- **Fine aggregate:** Locally available sand passing through 4.75mm sieve.
- **Coarse aggregate:** Locally available crushed blue granite stones aggregate of nominal size 10mm.
- **Water:** portable water that is good for concrete casting and curing available on site.
- **Concrete Admixture:** To improve workability of concrete, good performance and waterproofing.
- **Binding wire:** 16-gauge wire for tying reinforcement with each other.
- **Panel Mould:** As per the geometry and aesthetical requirements of compound wall Panel
- **Fibre:** To improve the strength of concrete
- **Mineral Admixtures:** Fly ash, Silica Fume, Rice Husk ash, Metakaoline, Ground granulated blast furnace slag etc or any other innovative admixtures which are easily available locally available material can be used for cost reduction or quality improvement of concrete.

Concrete and mineral admixtures are optional for improving the concrete quality and engineering properties. The proposed product can be manufactured with conventional concrete also.

DESCRIPTION OF THE DIAGRAM

Fig. 1: Flowchart of Methodology.

Fig. 2. Splitting of Bamboo in Strips

Fig. 3. Treatment on Strips of Bamboo with Coal Tar

Fig. 4: 3D view of precast prefab Bamboo Reinforced cement concrete compound walls panel

Fig. 5: Compound wall panel-Plan Elevations & Cross Sections

Fig. 6: Compound wall with precast concrete walls panels

Flowchart of Experimental Work

According the requirements of span, the geometry and strength of the compound wall panel will be engineered, for fixed sizes typical design can be adopted considering optimum span. The concrete mix design depends upon locally available material, ingredients and admixtures so that it should be cost-effective. The locally available bamboo should be well seasoned with the process of drying, splitting, bending, etc. To improve the performance, bamboo reinforcements should be coated with coal tar or epoxy as well as an anti-termite solution such as lethal solutions, etc.

The engineering properties of materials such as cement, aggregates, and bamboo reinforcement should be within the tolerable limit and maybe tested as per standard. As per the geometrical design, the mold should be prepared, and as per the aesthetic requirement, the inside face of the mold may be adopted with architectural design.

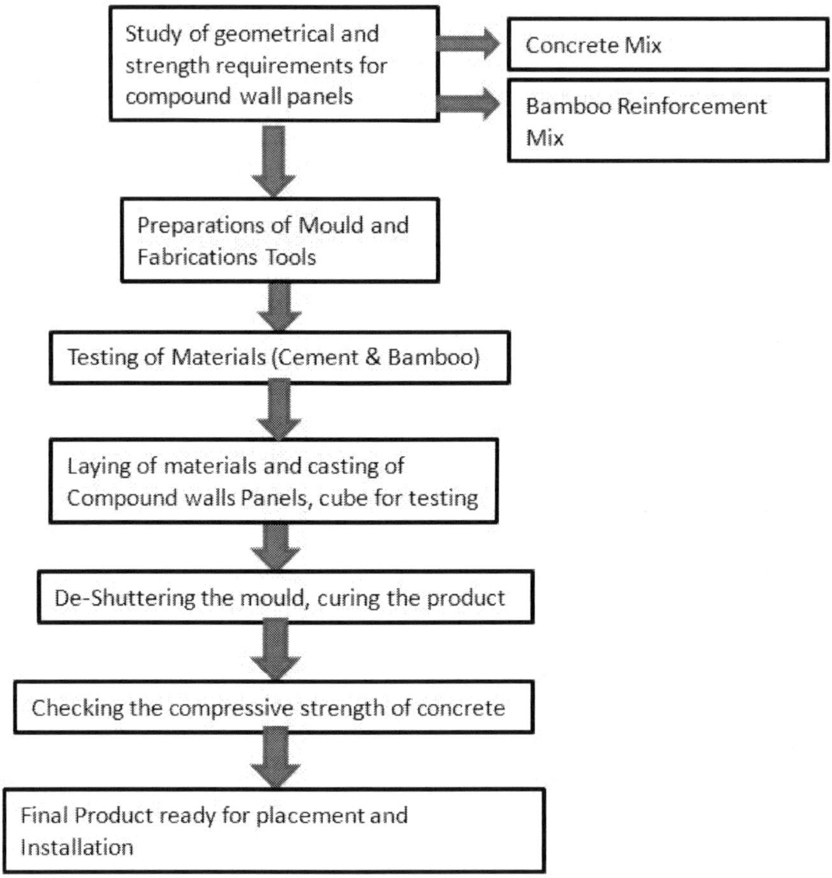

Figure 1: Flowchart of Methodology.

The mortar shall be poured into the mold with proper arrangement of bamboo reinforcements, needle vibrator or table vibrator should be used to remove voids, exposed surface of casting should be finished with equipment such funty or may provide any other texture design as per architectural or aesthetic requirements. The mortal sample may be taken and cast in the cube for testing of the concrete. The mold shall be de-shuttered after the setting of concrete and should be placed in curing yards. After the expected concreter test results the compound wall panels are ready to dispatch, placement and installations. These panels shall be installed in precast post having provision of groove to fix the compound wall panels. Using such type of bamboo reinforced compound wall panel , compound wall up to height of 5 m can be erected. The flowchart showing the methodology for the preparation process of precast prefab Bamboo Reinforced cement concrete compound walls panel is shown in Figure 1.

Seasoning of Bamboo

Seasoning is a process that will make bamboo compatible to use as reinforcement in concrete

- **Drying:** the harvested bamboo should be allowed to dry for three to four weeks before use to reduce the moisture content from the fibers
- **Splitting:** the bamboo can be done using a splitting machine or using a sharp knife to obtain straight sectional strips. Figure 2 shows the splitting of bamboo in strips.

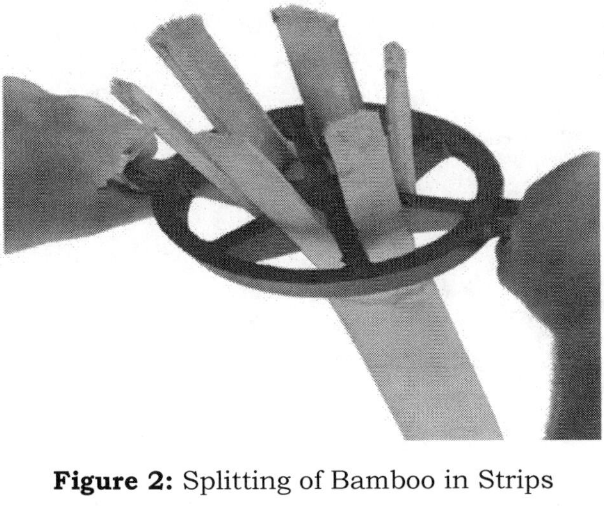

Figure 2: Splitting of Bamboo in Strips

- **Bending:** Bamboo strips can be bent with the application of pressure or heat. These bamboo strips can be used as stirrups for additional anchorage in the beam and column.

Treatment of Bamboo

To improve the performance of bamboo as reinforcement, the following treatments are to be carried to get optimal results. Figure 3 shows the treatment done on bamboo strips.

Figure 3: Treatment on Strips of Bamboo with Coal Tar

Waterproofing (Low water absorption): bamboo should be coated with coal tar epoxy paint for low water absorption as bamboo has the capacity to absorb up to 50% of the water of its weight from concrete.

Anti-termite: coating or soaking in lethal solution to resist termite attack.

Good bonding in concrete: a sprinkling of sand on the coated bamboo with coal tar and dried for 24-48 hours (Ahmad, 2014).

Dimensions

The compound wall Panel may be varies from 1500 to 3500mm in length 200 to 350mm in height and 50mm to 100 mm in width or thickness. The geometry should be decided as per requirement and structural design. 3D view of precast prefab Bamboo Reinforced cement concrete compound walls panel is shown in figure 4 along with possible variation in dimensions. Figure 5 shows plan elevations and cross sections of typical compound wall panel sections. Cross sections showing typical arrangements of longitudinal as well as vertical bamboo rebar. Photographs of Compound wall with precast concrete walls panels are shown in figure 6.

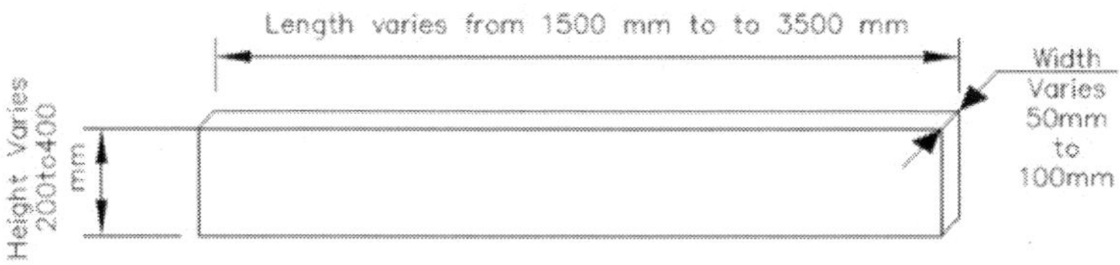

Figure 4: 3D view of precast prefab Bamboo Reinforced cement concrete compound walls panel

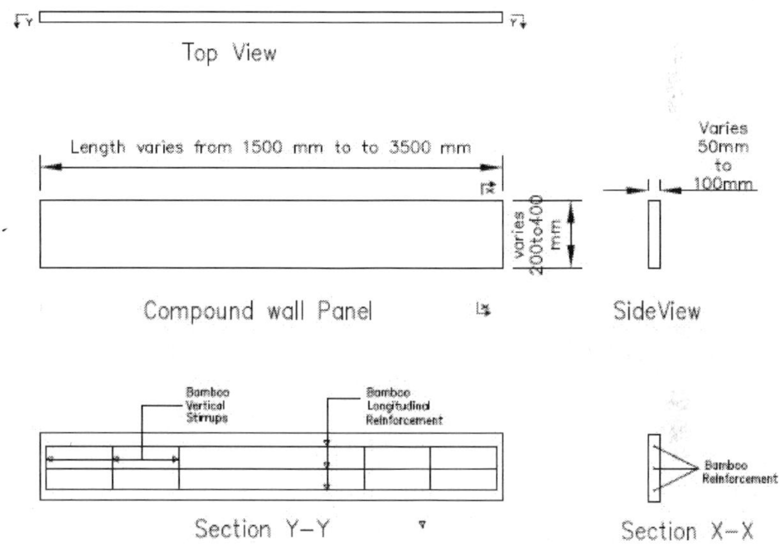

Figure 5: Compound wall panel-Plan Elevations & Cross Sections

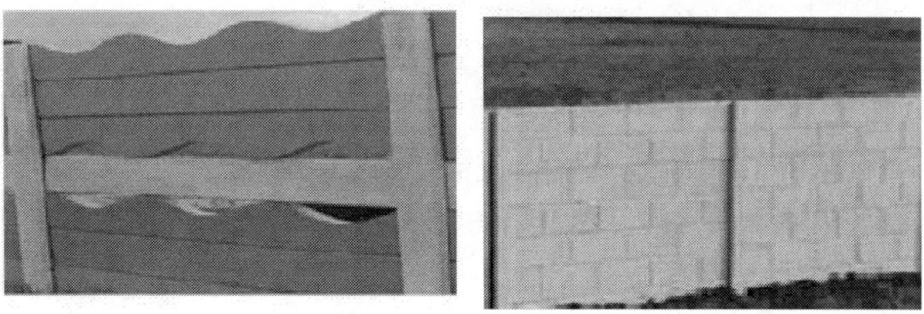

Figure 6: Compound wall with precast concrete walls panels

Casting of Precast Prefab Bamboo Reinforced Cement Concrete Compound Walls Panel

Precast prefab Bamboo Reinforced cement concrete compound walls panel would be cast on tables which are mounted on clamps fixed with rollers. As per the geometrical design, the mold should be prepared, and as per the aesthetic requirement, the inside face of the mold may be adopted with architectural design. Grade of concrete may vary from M15 ,M20,M25,M30. Concrete is then poured partly into the molds. Treated bamboo reinforcement shall be placed into the mold. Afterthis mold should be filled completely with concrete. After concrete pouring, elements are compacted through vibrators attached to the tables and finally finished. The mold shall be de-shuttered after the setting of concrete and should be placed in curing yards. After the expected concreter test results the compound wall panels are ready to dispatch, placement and installations.

PROBLEM STATEMENT

The major issues faced by the construction industry are maintaining quality as well as the speed of building constructions and the use of sustainable reinforcement. Modern-day builders and construction companies are forced to use precast concrete wall panels with steel reinforcement as there is no other alternative in the market. This research work proposes an approach to develop precast prefab Bamboo Reinforced cement concrete compound walls panels. This newly developed product is prefabricated, cheap, made from sustainable materials such as bamboo and fly ash, and has great potential in the construction industry.

WE CLAIMS

In Precast prefab concrete bamboo reinforced wall panels we are replacing traditional steel reinforcement with bamboo reinforcements. Also, we are allowing using locally available mineral admixtures such as fly ash, Silika Fume, Rice Husk ash, Metakaoline, Ground granulated blast furnace slag, etc or any other innovative admixtures for cost reduction or quality improvement of concrete. Due to precast products the quality of materials, curing of concrete, workmanship, speed of construction, and green-rated points can be archived with a higher degree.

- According to claim No.1 traditional steel reinforcements are being replaced with bamboo reinforcements which have the potential of getting similar results as steel reinforcement in the concrete according to many studies, which is being used in this product very effectively, economically, and also help to minimize carbon footprint.

- Also according to claim No.1 the mineral admixtures such as fly ash are easily available near thermal power plants, Similarly Rice Husk ash, Metakaoline, Ground granulated blast furnace slag are easily available in any part of the country which can be used very economically in this bamboo reinforced precast concrete wall panels and helps to achieve the green-sustainable product.
- As per claims No 2& 3 precast concrete bamboo reinforcement wall panels are very innovative, eco-friendly, fast, and easy of construction, ultimately cost-effective green product which will help society and industry to use natural products like bamboo and mineral admixtures.
- These Precast prefab concrete bamboo reinforced wall panels may also be use as portion wall panel for various industrial, residential, and commercial structures.

CITATIONS

- Mahatma Education Society, 202221029221, Mahatma Education Society, Chembur Naka, Mumbai - 400 071, Maharashtra, India. "Raster- Assistive AI based Book Reading Tool" 2022, Indian Patent Office.
- Mahatma Education Society, 202221028885, Mahatma Education Society, Chembur Naka, Mumbai - 400 071, Maharashtra, India." Increased the Soil Stability using W-Plastic." 2022, Indian Patent Office.
- Mahatma Education Society, 202221029340, Mahatma Education Society, Chembur Naka, Mumbai - 400 071, Maharashtra, India." Waste Kitchen and Plastic Convert into an Energy." 2022, Indian Patent Office.
- Mahatma Education Society, 202221029344, Mahatma Education Society, Chembur Naka, Mumbai - 400 071, Maharashtra, India." Method and Process to Improve the Basmati Rice Production", 2022, Indian Patent Office.
- Mahatma Education Society, 202241029342, Mahatma Education Society, Chembur Naka, Mumbai - 400 071, Maharashtra, India." Wireless Material Handling Management Method and Process", 2022, Indian Patent Office.
- Mahatma Education Society, 202221029339, Mahatma Education Society, Chembur Naka, Mumbai - 400 071, Maharashtra, India. "Design and Develop Alarm clock using AI and Cloud Data Based" 2022, Indian Patent Office.
- Mahatma Education Society, 202221029338, Mahatma Education Society, Chembur Naka, Mumbai - 400 071, Maharashtra, India, "Density Based Traffic Signals using Microcontroller using IoT- Based.", 2022, Indian Patent Office.
- Mahatma Education Society, 202221029337, Mahatma Education Society, Chembur Naka, Mumbai - 400 071, Maharashtra, India." Power Generation using Advanced Electromagnetic Suspension", 2022, Indian Patent Office.
- Nikhat Fatma Mumtaz Husain Shaikh, Dr. Farhat Fatma Mumtaz Husain Shaikh, 202221030181, Mahatma Education Society, Chembur Naka, Mumbai - 400 071, Maharashtra, India.," Refill Toothbrush with Slider" 2022, Indian Patent Office
- NAIR ROHAN UNNIKRISHNAN, 202221000050, MAHATMA EDUCATION SOCIETY'S, PILLAI COLLEGE OF ENGINEERING, Dr K M Vasudevan PILLAI Campus, Sector 16, New Panvel, NAVI Mumbai -410206, MH, India. "प्राबल्यम-ASCENDING" 2022, Indian Patent Office.
- Dr. Reni Francis,MAHATMA EDUCATION SOCIETY'S, PILLAI COLLEGE OF EDUCATION AND RESEARCH, 356107-001, Granted, "Math Teaching kit" 2022, Indian Patent Office.